Anatomy and Physiology
A Self-Instructional Course

Anatomy and Physiology
A Self-Instructional Course

3. The Locomotor System and The Special Senses

Written and designed by
Cambridge Communication Limited

Medical adviser

Bryan Broom MB BS(Lond)
General Practitioner
Beit Memorial Research Fellow
Middlesex Hospital Medical Research School

SECOND EDITION

Churchill Livingstone ⃟
EDINBURGH LONDON MELBOURNE AND NEW YORK 1985

CHURCHILL LIVINGSTONE
Medical Division of Longman Group Limited

Distributed in the United States of America by
Churchill Livingstone Inc., 650 Avenue of the
Americas, New York, N.Y. 10011, and by
associated companies, branches and
representatives throughout the world.

First edition 1977
Second edition 1985
  Reprinted 1991, 1993

ISBN 0-443-03207-6

British Library Cataloguing in Publication Data
A catalogue record for this book is available from
the British Library

Library of Congress Cataloging in Publication Data
Anatomy and physiology.
   Rev. ed. of: anatomy and physiology /
Ralph Rickards, David F. Chapman.   1977.
   Contents: 1. The human body and the reproductive
system — 2. The endocrine glands and the nervous
system — 3. The locomotor system and the special
senses — [etc.]
   1. Human physiology — Programmed instruction.
2. Anatomy, Human — Programmed instruction.
I. Broom, Bryan. II. Rickards, Ralph. Anatomy and
physiology. III. Cambridge Communication Limited.
QP34.5.A47 1984   612   84-4977

Printed in Hong Kong
WC/03

Contents

Contents

The Locomotor System

1. Bones and joints

1.1 The main parts of the skeleton

The adult skeleton consists of 206 bones. Bone is a living tissue with a rich blood and nerve supply. It can grow and repair itself after injury. Bone contains much crystalline inorganic material (mainly calcium salts) which makes it hard and rigid, but one-third of its substance is fibrous tissue which makes it tough and elastic.

The main functions of the bones of the skeleton are:

1. to act as a framework for the body, supporting it and giving it shape;

2. to provide a system of levers, moved by the action of muscles attached to them; to act as a system of levers, moved by the action of muscles attached to them;

3. to act as a reservoir of calcium, phosphorus, sodium and other elements;

4. to produce the red and white blood cells and platelets within the red marrow of certain bones.

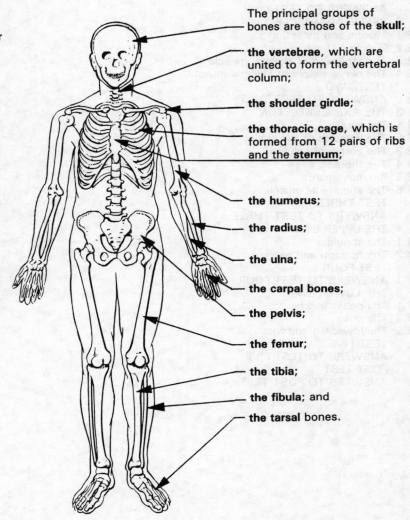

The principal groups of bones are those of the **skull**;

the vertebrae, which are united to form the vertebral column;

the shoulder girdle;

the thoracic cage, which is formed from 12 pairs of ribs and the **sternum**;

the humerus;

the radius;

the ulna;

the carpal bones;

the pelvis;

the femur;

the tibia;

the fibula; and

the tarsal bones.

1.2 The structure of bone

A typical long bone has a cylindrical shaft of hard bone, the diaphysis, containing a central cavity filled with **yellow marrow**.

The end of the bone, the *epiphysis*, consists of spongy bone covered by a thin layer of compact bone.

The epiphysis is coated with a smooth layer of *cartilage* where it moves against other bones.

Where there is no cartilage the bone is covered in **periosteum**. This is a sheet of fibrous tissue. It contains many blood vessels, and on its inner surface are cells capable of forming new bone, called *osteoblasts*.

The osteoblasts enable constant minor changes in bone structure to be made, and they initiate repair after injury.

All adult bone consists of thin sheets or *lamellae*. The bone cells, or *osteocytes*, lie on the surface of the lamellae and receive nourishment from the inner part of the bone.

Joint cartilage

Epiphysis

Diaphysis

Blood vessel

The lamellae are arranged in two ways:

1. In hard, or compact, bone the lamellae are arranged as concentric cylinders known as **Haversian systems**. These cylinders run along the length of the bone, and so are able to bear stress. Each cylinder has a central canal which contains blood vessels.

2. In spongy, or cancellous, bone the lamellae are arranged as a **honeycomb of irregular layers**. Red marrow, the blood forming tissue, or yellow marrow, a fatty tissue, occupies the spaces of the honeycomb network.

1.3 The development and growth of bone

The great majority of bones in the body develop from a cartilage model in the embryo. (A few urgently needed bones, e.g. the flat bones of the skull, develop from membraneous tissue. The end result is identical).

Development in a typical long bone

The bone is preceded by a cartilage model. ⎯⎯⎯⎯⎯⎯

A **periosteal collar** of new bone appears around the shaft of the model. The cartilage within the shaft becomes calcified. The cartilage cells die and leave spaces.

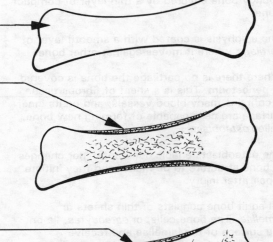

The honeycomb of degenerate cartilage is invaded by bone-forming cells, or *osteoblasts*, by blood vessels, and by bone-eroding cells, or *osteoclasts*. Bone is laid down in irregular layers within the cartilage model. ⎯⎯⎯⎯⎯

The ossification process extends down the shaft, and also begins separately at the epiphysis producing three centres of ossification. ⎯⎯⎯⎯⎯

Growth in the length of the bone occurs at the metaphysis, a sheet of healthy, living cartilage between the centres of ossification.

In the metaphysis the cartilage cells divide vertically. At first each cell produces healthy cartilage and expands pushing older cells down. Then the cells die. Then all the spaces enlarge to form vertical tunnels in the degenerating cartilage. These spaces are invaded by bone forming cells.

Growth in length stops in adult life when the epiphysis fuses with the shaft.

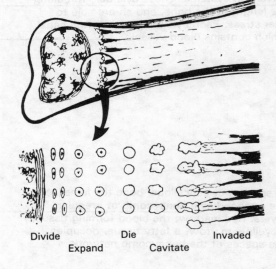

Divide		Die		Invaded
	Expand		Cavitate	

During growth bones are continually altering in shape to meet altering forces and to accommodate developing organs.

Repair of bones after injury is largely due to the activity of the osteoblasts beneath the periosteum.

1.4 The joints

Articulations, or joints, are the junctions between two adjacent bones. They are classified according to their structure.

1. Fibrous joints

The bones are joined by tough collagen fibres. The joint is usually fixed e.g. the **sutures of the skull**. Occasionally there may be some slight movement.

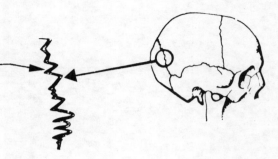

2. Cartilaginous joints

The bone surfaces are covered by a layer of cartilage and joined by tough fibrous tissue embedded into the cartilage, e.g. between the bodies of the **vertebrae**, and the **pubic symphysis**.

These joints usually allow a small degree of movement.

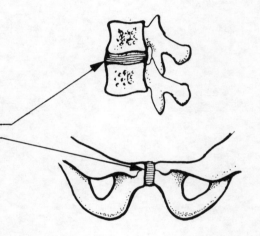

3. Synovial joints

These are the commonest type of joint. They usually allow a great degree of movement (e.g. knee, shoulder, elbow, wrist etc.) but some synovial joints are relatively immobile (e.g. the sacro-iliac joint). The joint is enclosed in a **fibrous capsule**, lined with a thin **synovial membrane**. This membrane secretes **synovial fluid** into the joint space to lubricate the joint. The surfaces of the bone are lined with smooth, hard-wearing articular **cartilage** where they come into contact with other bones.

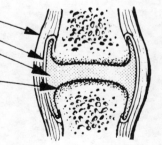

In some joints there is a crescent of fibrous cartilage which partly separates the bones of the joint (e.g. the knee, the jaw).

Synovial joints are prevented from dislocating by the surrounding muscles which reflexly prevent extreme movements. They are also supported by strong ligaments. Many ligaments are thickenings of the fibrous joint capsule.

TEST ONE

ANSWERS TO TEST ONE

1. (a)   Which cells enable the formation of new bone, structural changes in bones and the repair of injury?

   _____

   (b)   Where are these cells located?

   _____

2. (a)   How many pairs of ribs are there?

   (b)   How many long bones are there between the knee and the ankle?

   (c)   Name two elements which are stored in bone.

   _____

3. What is the correct name for the shaft of a growing bone?

   (a)   Epiphysis.

   (b)   Diaphysis.

   (c)   Metaphysis.

4. Which of the statements on the right below apply to the items listed on the left?

   (i) Haversian systems.   (a)   The fibrous and cellular layer clothing bone.
   (ii) Osteoblasts.   (b)   Microscopic cylinders of bone tissue.
   (iii) Periosteum.   (c)   The region where bone growth occurs.
   (iv) Metaphysis.   (d)   Cells which lay down bone.

5. (a)   At what part of a long bone does growth in length occur?

   _____

   (b)   What happens to the bone in adulthood that stops further bone growth?

   _____

6. (a)   Name the three types of joint.

   (b)   Which is the most common?

   _____

   _____

ANSWERS TO TEST ONE

1. (a) Osteoblasts.
(b) On the inner surface of the periosteum.

2. (a) Twelve.
(b) Two.
(c) Two of: sodium, phosphorus, calcium.

3. (b) Diaphysis.

4. (i) Haversian systems   (b)   are microscopic cylinders of bone tissue.
(ii) Osteoblasts   (d)   are cells which lay down bone.
(iii) Periosteum   (a)   is the fibrous and cellular layer clothing bone.
(iv) Metaphysis   (c)   is the region where bone growth occurs.

5. (a) The metaphysis is the site for growth in the length of a bone.
(b) The fusion of the epiphysis with the shaft prevents further growth of the bone.

6. (a) Fibrous, cartilaginous, synovial.

(b) Synovial.

2. Skeletal muscle

2.1 Muscle and its action

Skeletal muscle accounts for half the body weight of a typical adult. Its primary function is to move bones at their articulations. It does this by shortening (contracting). By lengthening (relaxing), muscles allow other muscles to contract and move bones.

Muscle may be attached directly to bone, but where its bulk would interfere with function (e.g. in the hand), where it would rub over bone, or where its action needs to be concentrated, it is attached by white fibrous tendons. Tendons may be like cords, like straps, or even like sheets (e.g. at the front of the abdomen).

No muscle acts alone. It always acts as part of a group under the control of the nervous system.

Muscle function can be illustrated by considering the upper arm.

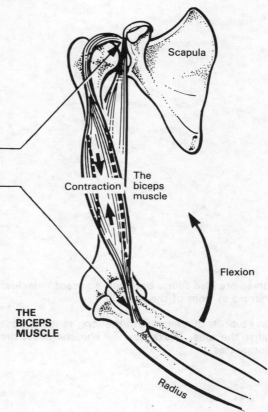

The biceps muscle of the upper arm is attached by a tendon to the scapula. This attachment usually remains stationary and is the **origin** of the muscle. The other end of the muscle is attached to the radius. This attachment is to the mobile bone and is known as the **insertion** of the muscle.

The biceps is a flexor muscle; it bends the joint, raising the forearm as it shortens. It also tends to roll the forearm over to bring it palm upward because of its point of insertion.

The triceps muscle on the back of the upper arm is an extensor muscle; it straightens the joint, having the opposite action to the biceps.

During simple flexion (bending) of the elbow therefore:

  (i)   the biceps contracts — it is the *prime mover*;
  (ii) the triceps relaxes reflexly — it is the *antagonist*;
  (iii) certain muscles in the forearm contract to prevent the rolling-over movement;
  (iv) muscles around the shoulder contract to steady the shoulder joint.

} they are the *synergists*

2.2 Bursae

Bursae are thin-walled sacs of synovial membrane containing a little synovial fluid. They are found between muscle and bone at points where the muscle moves over the bone. They allow for free-gliding movement of the muscle, the synovial membrane acting as a lubricant.

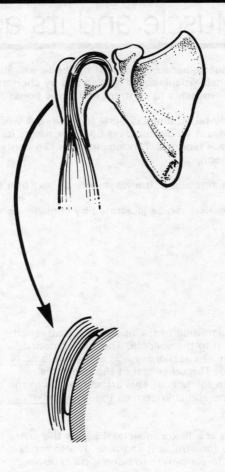

Bursae are also found between adjacent muscles, and beneath the skin surface over certain joints, for example in front of the knee.

Long tubular bursae sheath the long tendons passing from the forearm to the fingers, and from the calf to the toes. These *synovial sheaths* allow the tendons to move freely over a range of several centimetres.

2.3 The structure of skeletal muscle

Skeletal muscle is made up of large numbers of muscle fibres. These are extremely long, unbranched **cylindrical cells**. They are supported by connective tissue and have a rich blood and nerve supply.

Each cell has many **nuclei** and has a striped appearance. Its **wall**, or sarcolemma, contains **myofibrils** closely packed in a liquid sarcoplasm. There are also many mitochondria.

The red colour of muscle is due to myoglobin a haemoglobin-like protein in the sarcoplasm.

Each myofibril has alternate light and dark striations, called I and A bands respectively. The striations are caused by two types of filament, one consisting of the protein **actin**, and the other consisting of the protein **myosin**.

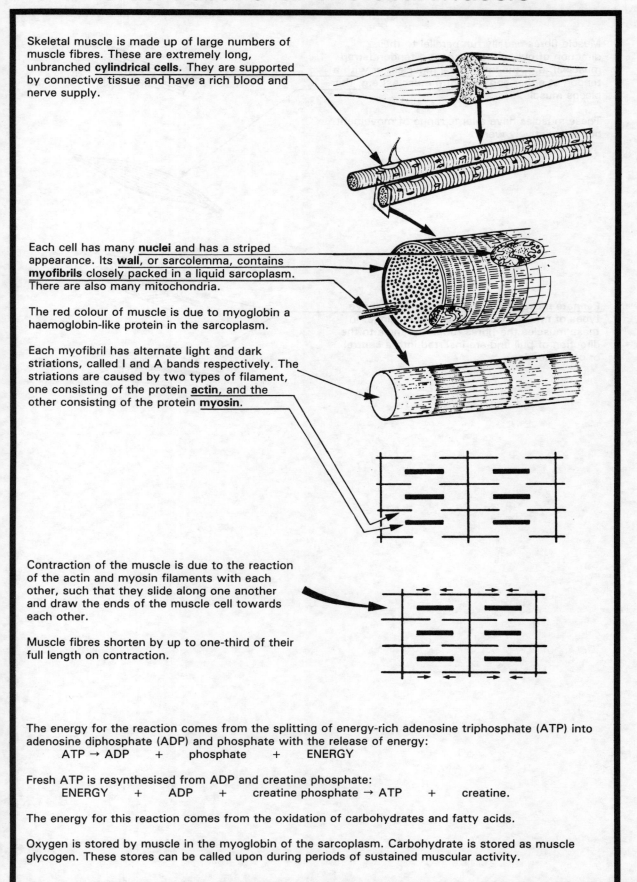

Contraction of the muscle is due to the reaction of the actin and myosin filaments with each other, such that they slide along one another and draw the ends of the muscle cell towards each other.

Muscle fibres shorten by up to one-third of their full length on contraction.

The energy for the reaction comes from the splitting of energy-rich adenosine triphosphate (ATP) into adenosine diphosphate (ADP) and phosphate with the release of energy:

ATP → ADP + phosphate + ENERGY

Fresh ATP is resynthesised from ADP and creatine phosphate:

ENERGY + ADP + creatine phosphate → ATP + creatine.

The energy for this reaction comes from the oxidation of carbohydrates and fatty acids.

Oxygen is stored by muscle in the myoglobin of the sarcoplasm. Carbohydrate is stored as muscle glycogen. These stores can be called upon during periods of sustained muscular activity.

Muscle fibres usually run parallel to their
direction of pull, either without a tendon (**strap
muscles**), e.g. the intercostal muscles, or with a
tendon at either end (**fusiform muscles**) e.g. the
biceps muscle.

These muscles have a large range of movement
but are relatively weak.

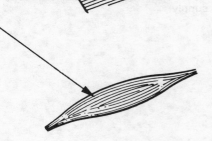

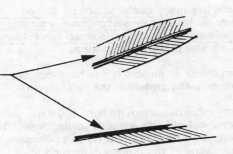

Pennate muscles are stronger than the above
types of muscle, but have a shorter range. In
these muscles the fibres run at an angle to the
direction of pull and are inserted into a central
or offset tendon.

2.4 The nerve supply of skeletal muscle

Muscles are supplied by two sorts of nerve fibre:

1. **sensory nerves** which carry impulses from the muscle, mainly from specialised stretch receptors, the muscle spindles;

2. **motor nerves** which carry impulses to the muscle fibres to trigger their contraction.

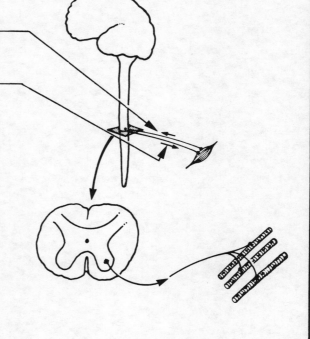

The cell bodies of the motor nerve cells lie in the anterior horn of grey matter in the spinal cord.

Each nerve cell has a main fibre, or axon, which branches to supply from 50 to 2000 muscle fibres. The fineness of the movement of the muscle depends on the number of fibres supplied by each nerve.

All the cell bodies supplying one muscle lie close together in the spinal cord.

Nerve impulses reach each muscle fibre at about its centre, at a motor end plate. The incoming nerve impulse causes stored acetylcholine to be released from the **motor end plate**. The acetylcholine acts to amplify the nerve impulse. It causes a large wave of electrical activity to sweep along the muscle, initiating changes which cause the muscle to **contract**.

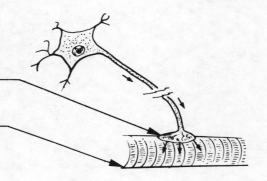

The strength of the contraction depends on the number of fibres stimulated. When the impulses stop the muscle relaxes.

TEST TWO

1. **Which of the terms on the left apply to the items on the right, in relation to simple flexion of the elbow?**

   (i)   Origin.

   (ii) Insertion.

   (iii) Prime mover.

   (iv) Antagonist.

   (v)   Synergist.

   (a) Triceps.

   (b) Shoulder muscles.

   (c) Radius.

   (d) Biceps.

   (e) Scapula.

2. **Among the following types of muscles — strap muscles, fusiform muscles, pennate muscles — one type forms an exception; which is it and why?**

   _____

   _____

3. **Indicate which of the names in the list below refer to the parts of the muscle fibre labelled on the diagram alongside by placing the appropriate letters in the brackets.**

   1. Sarcolemma.   (   )
   2. End plate.   (   )
   3. Nucleus.   (   )
   4. Motor nerve.   (   )
   5. Myofibril.   (   )

4. **Place ticks in the appropriate brackets.**

	Muscle spindle	Motor end plate	Anterior horn cell
(a) Sensory nerve impulses come from here.	( )	( )	( )
(b) Motor nerve impulses come from here.	( )	( )	( )
(c) Motor nerve impulses go to here.	( )	( )	( )

5. **On what does the strength of contraction of a muscle depend?**

   _____

   _____

6. **Where is the acetylcholine which causes muscle contraction stored?**

   _____

   _____

ANSWERS TO TEST TWO

1. (i)   Origin.   (e)   Scapula.

    (ii) Insertion.   (c)   Radius.

    (iii) Prime mover.   (d)   Biceps.

    (iv) Antagonist.   (a)   Triceps.

    (v)   Synergist.   (b)   Shoulder muscles.

2. Unlike the other two types, strap muscles are not attached to tendons. Fusiform muscles have tendons at either end, and pennate muscles are inserted into a central or offset tendon.

3. 1. Sarcolemma.   (C)
    2. End plate.   (B)
    3. Nucleus.   (E)
    4. Motor nerve.   (A)
    5. Myofibril.   (D)

4.

		Muscle spindle	Motor end plate	Anterior horn cell
(a)	Sensory nerve impulses come from here	(√)	( )	( )
(b)	Motor nerve impulses come from here	( )	( )	(√)
(c)	Motor nerve impulses go to here	( )	(√)	( )

5. On the number of fibres stimulated.

6. In the motor end plate.

3. The axial skeleton

3.1 The skull

Apart from the lower jaw or mandible, and the ear ossicles, the individual bones of the skull are firmly joined to each other by fibrous joints or sutures. They form a series of bony cavities.

The cranial cavity contains the brain and its coverings. Its sloping floor, the base of the skull, contains three hollows known as the cranial fossae. There is a large opening, the foramen magnum, in the posterior fossa through which the spinal cord passes. The facial bones are suspended from the front part of the base of the skull and form the two orbital cavities and the nasal and oral cavities.

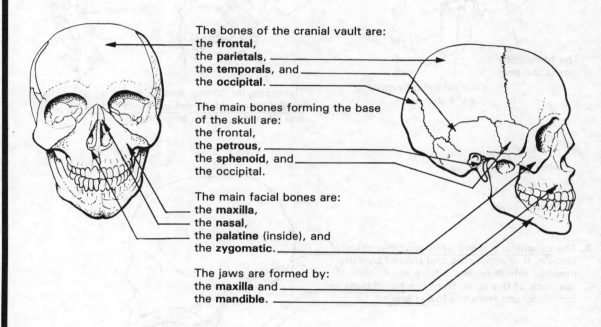

The bones of the cranial vault are:
the **frontal**,
the **parietals**,
the **temporals**, and
the **occipital**.

The main bones forming the base of the skull are:
the frontal,
the **petrous**,
the **sphenoid**, and
the occipital.

The main facial bones are:
the **maxilla**,
the **nasal**,
the **palatine** (inside), and
the **zygomatic**.

The jaws are formed by:
the **maxilla** and
the **mandible**.

A vertical cross section through the skull shows:

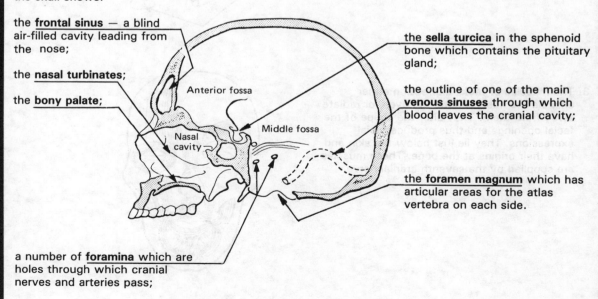

the **frontal sinus** — a blind air-filled cavity leading from the nose;

the **nasal turbinates**;

the **bony palate**;

Anterior fossa

Nasal cavity

Middle fossa

the **sella turcica** in the sphenoid bone which contains the pituitary gland;

the outline of one of the main **venous sinuses** through which blood leaves the cranial cavity;

the **foramen magnum** which has articular areas for the atlas vertebra on each side.

a number of **foramina** which are holes through which cranial nerves and arteries pass;

3.2 The muscles of the skull

1. The muscles of mastication are bulky and powerful. They are supplied by the fifth cranial nerve.

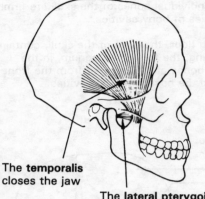

The **temporalis** closes the jaw

The **lateral pterygoid** opens the jaw.

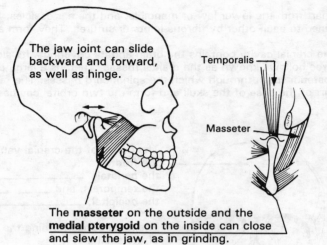

The jaw joint can slide backward and forward, as well as hinge.

Temporalis

Masseter

The **masseter** on the outside and the **medial pterygoid** on the inside can close and slew the jaw, as in grinding.

2. The tongue is a mass of interwoven skeletal muscle. It is supported and moved by other muscles which originate from either side of the base of the skull, from the **hyoid bone** in the neck, and from the **lower jaw**.

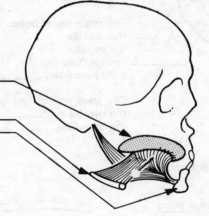

3. The muscles of facial expression either encircle the nose, mouth and eyes, or radiate away from them. They alter the shape of the facial openings and thus produce facial expressions. They lie just below the skin and have their origins at the bone. These muscles are supplied by the seventh cranial nerve.

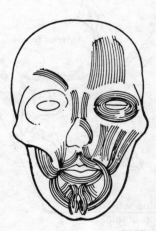

4. The pharynx is the hollow muscular tube which forms the back of the mouth, the nose, and larynx. It is open at its front and is lined inside with mucous membrane.

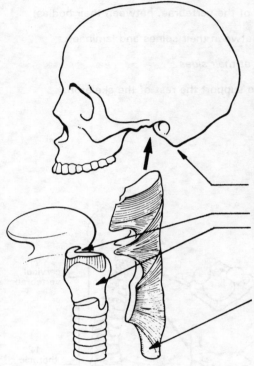

The pharynx is a funnel carrying both food and air. Its muscle is used to propel food downward during the second stage of swallowing.

It lies just in front of the vertebral column.

It is attached:
to **the base of the skull**,
to the muscle in the wall of the cheek,
to **the hyoid bone**, and
to **the larynx**.

It leads directly into **the oesophagus** and **the larynx**.

5. The major muscles of the neck are the following:

The **cervical vertebral muscles** keep the head erect. They form a complex, massive group around the cervical spine.

The **sternocleidomastoid muscle** is large and obvious. It pulls the head forward and turns it to the side opposite itself.

The **anterior strap muscles** run from the chin via the **hyoid** and **larynx** to the sternum. They help form the floor of the mouth, they support the larynx, and they act to flex the neck.

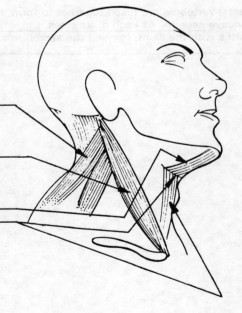

3.3 The vertebral column

The vertebral column is made up of 33 irregular bones, the vertebrae. The vertebrae encircle the spinal cord.

Vertebrae are attached to adjoining vertebrae by:

(i)　intervertebral discs of fibrocartilage *at the front* of the vertebrae, between their bodies;

(ii)　strong ligaments *at the back* of the vertebrae, between their spines and laminae;

(iii)　synovial joints, between the articular processes *at their sides*.

The vertebrae thus form a strong but flexible structure to support the rest of the skeleton.

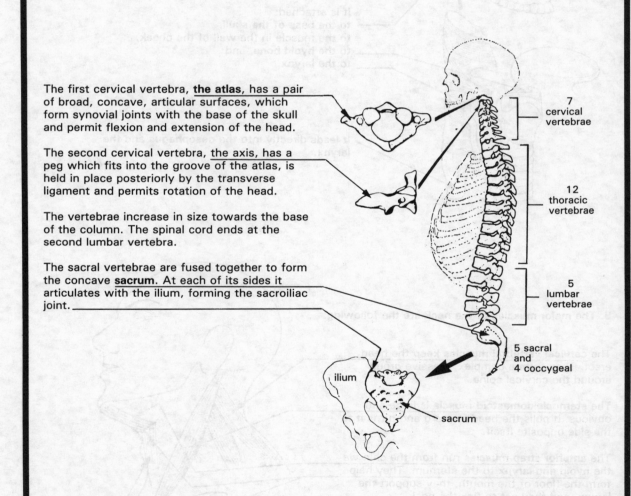

The first cervical vertebra, **the atlas,** has a pair of broad, concave, articular surfaces, which form synovial joints with the base of the skull and permit flexion and extension of the head.

The second cervical vertebra, the axis, has a peg which fits into the groove of the atlas, is held in place posteriorly by the transverse ligament and permits rotation of the head.

The vertebrae increase in size towards the base of the column. The spinal cord ends at the second lumbar vertebra.

The sacral vertebrae are fused together to form the concave **sacrum**. At each of its sides it articulates with the ilium, forming the sacroiliac joint.

7 cervical vertebrae

12 thoracic vertebrae

5 lumbar vertebrae

5 sacral and 4 coccygeal

ilium

sacrum

The main features of a vertebra can be illustrated by reference to one of the thoracic vertebrae. There are, however, characteristic differences between the cervical, thoracic, and lumbar vertebrae.

Viewed from above each vertebra can be seen to have:

a **dorsal spinous process** for the attachment of muscle;

two pairs of **articular processes** (superior and inferor),

a **spinal canal**,

a **body**,
which is the solid, cylindrical, weight-bearing part of the vertebra containing red marrow,

two **rounded pedicles**

two **flat roof-like laminae**

two **transverse processes** for muscle attachment (in the thoracic vertebrae each transverse process has an articulation for a rib.) The lamina and pedicles form the neural arch.

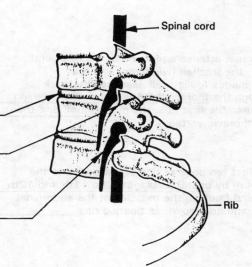

A thoracic vertebra

Rib

Spinal cord

Viewed from the side:

each vertebra can be seen to be separated from the next by a thick ring of fibro cartilage. This **intervertebral disc** has a fluid centre and acts as a shock absorber.

Synovial joints lie between the inferior articular process of one vertebra and the superior articular process of the one below. These joints limit the mobility of the vertebral column.

Spinal nerves leave the cord between the pedicles, through the *intervertebral foramina*.

Rib

A mass of muscles lies alongside the vertebral column, particularly posteriorly, and forms the main extensors. The muscles on the front of the abdomen are the most powerful flexors.

3.4 The thoracic cage

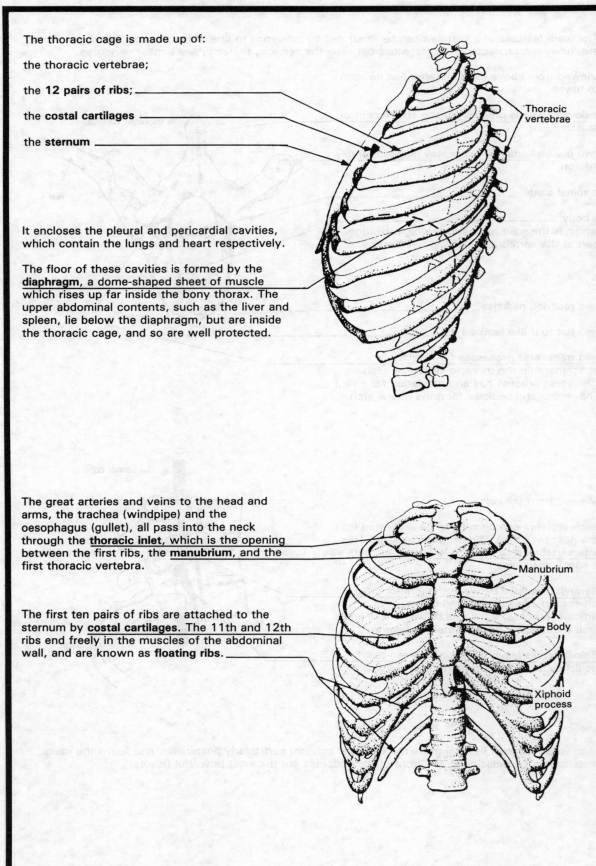

The thoracic cage is made up of:

the thoracic vertebrae;

the **12 pairs of ribs;**

the **costal cartilages**

the **sternum**

Thoracic vertebrae

It encloses the pleural and pericardial cavities, which contain the lungs and heart respectively.

The floor of these cavities is formed by the **diaphragm**, a dome-shaped sheet of muscle which rises up far inside the bony thorax. The upper abdominal contents, such as the liver and spleen, lie below the diaphragm, but are inside the thoracic cage, and so are well protected.

The great arteries and veins to the head and arms, the trachea (windpipe) and the oesophagus (gullet), all pass into the neck through the **thoracic inlet**, which is the opening between the first ribs, the **manubrium**, and the first thoracic vertebra.

The first ten pairs of ribs are attached to the sternum by **costal cartilages**. The 11th and 12th ribs end freely in the muscles of the abdominal wall, and are known as **floating ribs**.

Manubrium

Body

Xiphoid process

3.5 Rib movements

When viewed from the side, most of the ribs can be seen to droop down a little below a line joining their mobile ends.

The spaces between the ribs are filled with muscle, the intercostal muscles. Contraction of these muscles pulls the ribs upwards and outwards for inspiration, while relaxation of the intercostal muscles allows the ribs to move downwards and inwards for expiration.

Since the bones around the thoracic inlet are relatively fixed, contraction of the *intercostal muscles* draws the ribs upward.

This movement causes the ribs to swing outward a little, because of their drooping shape. The diameter and volume of the chest increases and causes air to be drawn into the lungs.

This respiratory action is the main function of the ribs. Their protective action is secondary.

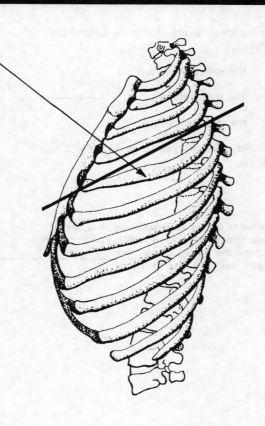

3.6 The abdominal muscles

There are three layers of muscle around the abdomen:

> *the external oblique*
>
> *the transversus*
>
> *the internal oblique*

The fibres of each layer run in a different direction.

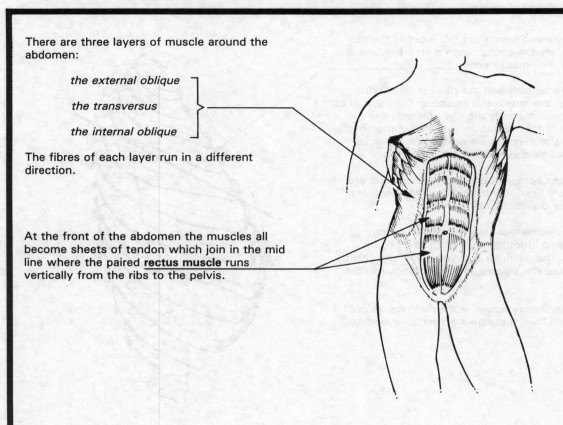

At the front of the abdomen the muscles all become sheets of tendon which join in the mid line where the paired **rectus muscle** runs vertically from the ribs to the pelvis.

The layered muscles arise from the lower ribs, the sides of the lumbar vertebrae, and the brim of the pelvis. They restrain and protect the abdominal contents. Their contraction raises the pressure inside the abdomen thus assisting micturition, defecation, and childbirth.

The rectus muscle is a powerful flexor of the lumbar vertebral column.

TEST THREE

ANSWERS TO TEST THREE

1. (a)   What does the cranial cavity contain?   _____

   (b)   What passes through the foramen magnum?   _____

   (c)   What lies in the sella turcica?   _____

   (d)   Where are the turbinates?   _____

2. Label six major bones seen here.
   Tick one that is part of the base of the skull.

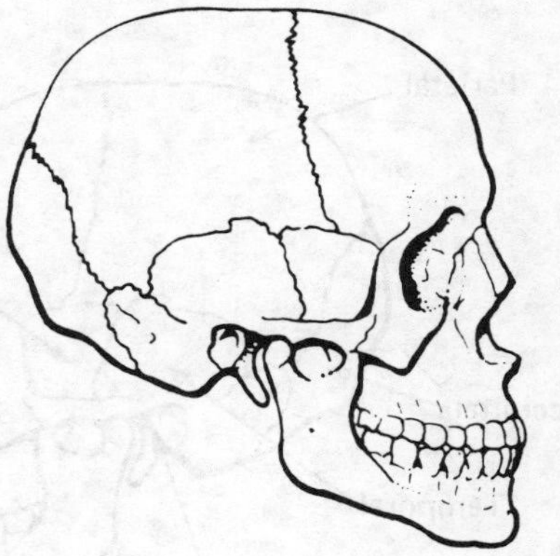

3. (a)   What is the name of this muscle?

   _____

   (b)   Does it turn the head towards, or away from, itself?

   _____

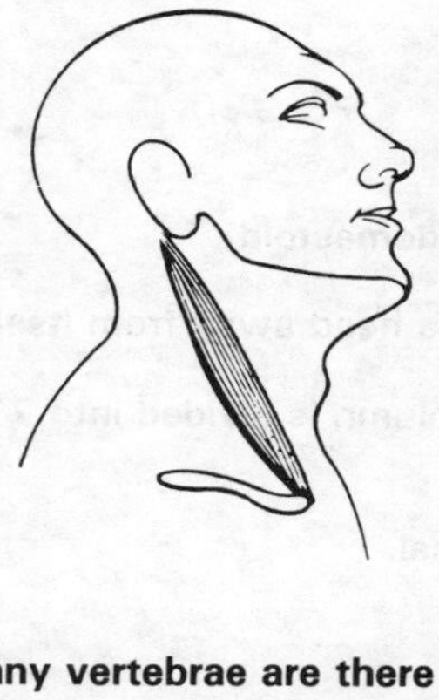

4. What are the divisions of the vertebral column, and how many vertebrae are there in each division?

   _____

   _____

5. Indicate which of the names in the list below refer to the parts of the vertebra labelled on the diagram alongside by placing the appropriate letters in the brackets.

   (i)   Spinal canal.   (   )

   (ii)   Lamina.   (   )

   (iii) Transverse process.   (   )

   (iv)   Pedicle.   (   )

   (v)   Body.   (   )

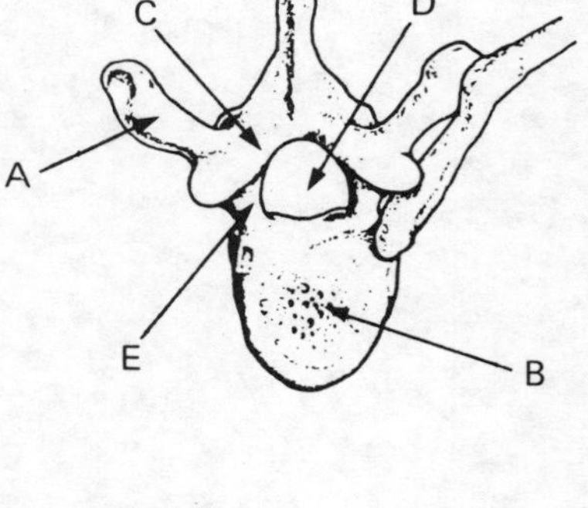

ANSWERS TO TEST THREE

1. (a)   The brain (and its coverings).

   (b)   The spinal cord.

   (c)   The pituitary gland.

   (d)   In the nasal cavity.

2. Six of:

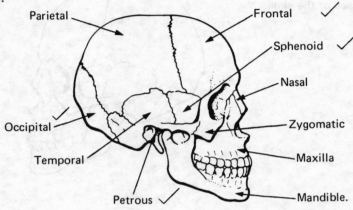

3. (a)   Sternocleidomastoid.

   (b)   It turns the head away from itself.

4. The vertebral column is divided into 7 cervical, 12 thoracic, 5 lumbar, 5 sacral and 4 coccygeal vertebrae.

5. (i)   Spinal canal.   (D)

    (ii)   Lamina.   (C)

    (iii) Transverse process.   (A)

    (iv)   Pedicle.   (E)

    (v)   Body.   (B)

4. The upper limb

4.1 The shoulder

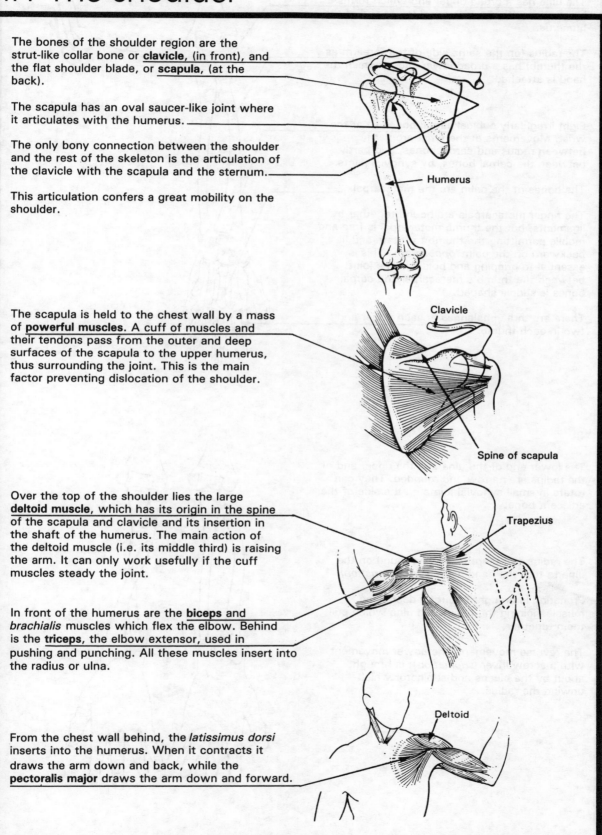

The bones of the shoulder region are the strut-like collar bone or **clavicle**, (in front), and the flat shoulder blade, or **scapula**, (at the back).

The scapula has an oval saucer-like joint where it articulates with the humerus.

The only bony connection between the shoulder and the rest of the skeleton is the articulation of the clavicle with the scapula and the sternum.

This articulation confers a great mobility on the shoulder.

The scapula is held to the chest wall by a mass of **powerful muscles**. A cuff of muscles and their tendons pass from the outer and deep surfaces of the scapula to the upper humerus, thus surrounding the joint. This is the main factor preventing dislocation of the shoulder.

Over the top of the shoulder lies the large **deltoid muscle**, which has its origin in the spine of the scapula and clavicle and its insertion in the shaft of the humerus. The main action of the deltoid muscle (i.e. its middle third) is raising the arm. It can only work usefully if the cuff muscles steady the joint.

In front of the humerus are the **biceps** and *brachialis* muscles which flex the elbow. Behind is the **triceps**, the elbow extensor, used in pushing and punching. All these muscles insert into the radius or ulna.

From the chest wall behind, the *latissimus dorsi* inserts into the humerus. When it contracts it draws the arm down and back, while the **pectoralis major** draws the arm down and forward.

Humerus

Clavicle

Spine of scapula

Trapezius

Deltoid

4.2 The forearm and hand

The ulna has a broad upper end which forms a hinge joint with the cylindrical lower end of the humerus.

The radius (on the same side of the forearm as the thumb) has a broad lower end to which the hand is attached.

Triceps tendon

Eight irregularly cubical **carpal bones** lie at the wrist. Movements at the wrist are partly between radius and carpal bones, and partly between the carpal bones, by synovial joints.

The bones of the palm are the **metacarpals**.

The finger metacarpals are bound together by ligaments, but the thumb metacarpal is free and mobile permitting the thumb to rotate and lie backward on the palm *(opposition)*. This is essential to gripping and holding. The joint between the thumb's metacarpal and carpal bones is saddle shaped.

There are three **phalanges** in each finger and two in each thumb.

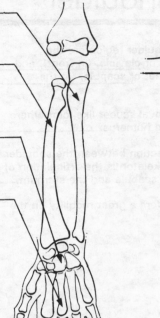

The lower end of the ulna and the upper end of the radius are narrow and rounded. They can rotate in small synovial joints on the side of the adjacent bone.

The radius can therefore swing round on the ulna to turn the hand palm down *(pronation)*.

Pronation is brought about by deep-lying muscles joining the radius and ulna which pull them together.

The reverse movement, the power movement with a screwdriver *(supination)*, is brought about by the biceps and supinator which unwind the radius.

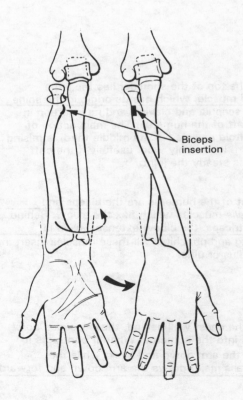

Biceps insertion

The bulk of the muscles of the forearm are inserted by strong cord-like tendons into the carpal and terminal bones (phalanges) of the fingers.

The powerful flexors, for gripping, lie in layers on the palm side of the forearm.

The weaker extensors lie on the dorsum. As well as straightening the fingers and wrist, these muscles are essential for a strong grip as the wrist must first be extended and fixed to take up slack in the flexor tendons.

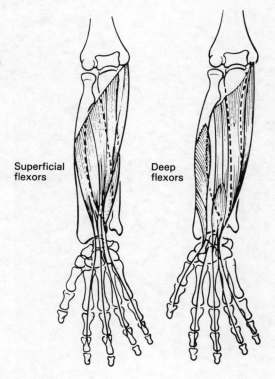

Superficial flexors

Deep flexors

Movements at the wrist joint involve mainly the radial and ulnar flexors and the extensors of the carpus.

Where the tendons cross the wrist they are prevented from bow stringing by passing through the **carpal tunnel** which is formed by strong restraining transverse fibrous bands anchored at each end to the bone.

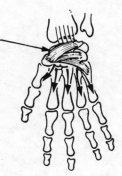

20 small muscles lie between the metacarpals and in the flesh of the palm, including the ball of the thumb. These muscles spread and close the fingers, and rotate the thumb. They are small, but modify the action of the forearm muscles and are essential for fine hand movements.

Some long forearm muscles insert into the bases of the index and metacarpal bones and move, and steady, the wrist.

TEST FOUR

ANSWERS TO TEST FOUR

1. In what actions are the following muscles used?

(a) Brachialis muscles.
(b) Triceps.
(c) Deltoid muscle.
(d) Biceps.

2. Which of the statements on the right below apply to the bones listed on the left?

(i) The radius.
(ii) The ulna.
(iii) The humerus.

(a) Has a broad upper end.
(b) Has a narrow rounded upper end.
(c) Has a cylindrical lower end.

3. Which muscle acts as a supinator of the hand?

_____

4. Which bones lie in the palm?

_____

5. (a) Name the features indicated on the diagram below.

(b) Is this a view of the thumb side or the little finger side of the elbow (palm up)?

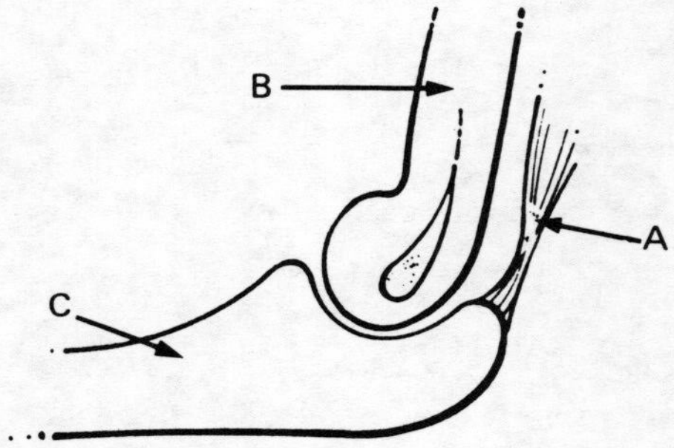

6. What prevents the tendons at the wrist from popping out when the wrist is bent upwards?

_____

ANSWERS TO TEST FOUR

1. (a)   The Brachialis muscles and (d) the biceps are used in flexing the elbow.
   (b)   The triceps are used in extending the elbow.
   (c)   The deltoid muscles are used in raising the arm.

2.   (i)   The radius   (b)   has a narrow rounded upper end.
     (ii) The ulna   (a)   has a broad upper end.
     (iii) The humerus   (c)   has a cylindrical lower end.

3. The biceps.

4. The metacarpals.

5. (a)   A. Triceps muscle.
         B. Humerus.
         C. Ulna.

   (b)   The little finger side.

6. The carpal tunnel.

5. The lower limb

5.1 The pelvis and hip

The pelvis is a basin with a wide brim at either side.

It is formed by two pelvic bones which join posteriorly to the *sacrum* and anteriorly at the *pubic symphysis*.

The pelvic basin contains the lower abdominal contents. It is wider in the female than in the male, for child bearing.

The pelvic bone has three fused parts:

the *ilium*;

the *ischium* (on which we sit);

the *pubis* (at the front).

These join at the **acetabulum**, the socket of the hip joint.

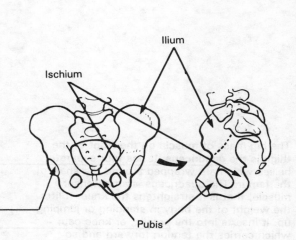

The *femur* is the largest and thickest bone in the body. It has a ball-shaped head, which fits into the acetabulum, on the end of a stout neck.

The **psoas muscle**, from the front of the lumbar vertebrae, joins with the **iliacus muscle**, skims the pelvis basin and inserts into the femur. It flexes the hip, bringing the thigh up.

The **gluteal muscles** arise from the back of the ilium and insert, via a tendon, into the greater trochanter of the femur. Their main function is to tilt the pelvis and raise the legs alternatively in walking. They are covered behind by the *gluteus maximus,* the buttock muscle, which is an extensor of the hip and brings the trunk into the upright position.

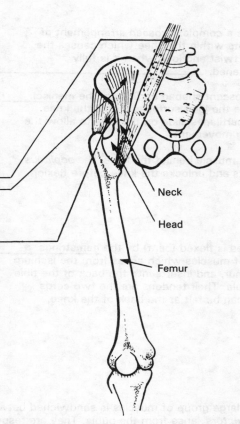

Many short thick muscles arise from the pelvis and insert into the upper femur to steady the hip joint. The joint possesses strong spiral ligaments which prevent backward extension of the hip.

5.2 The knee

At the knee joint two **condyles** at the lower end of the femur roll over the flat upper end of the **tibia**. Strong ligaments support the joint on either side.

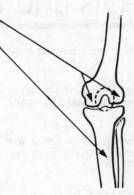

The big mass of muscle on the front of the thigh is the **quadriceps**. It has four separate bellies which are wrapped around the shaft of the femur. The quadriceps is an extensor muscle; that is it straightens the knee. It lifts the weight of the body in standing or jumping up. It inserts into the **patella**, or knee cap, which carries the tendon forward and so improves the efficiency of the joint. It then inserts into the front of the tibia via the **patellar tendon**.

Joint ligaments

There is a complex crossed arrangement of ligaments within the knee which causes the joint to twist and lock when it is fully straightened.

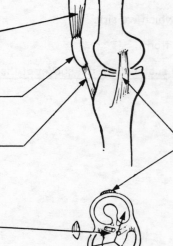

Two crescent-shaped cartilages, the **menisci**, separate the joint surfaces within the knee. These cartilages slide on the tibia to allow the twisting movement.

A small muscle behind the knee, the *popliteus*, untwists and unlocks the knee before flexion.

The knee is flexed (bent) by the **hamstrings**, a group of muscles which arises from the ischium and femur, and inserts into the back of the tibia and fibula. Their tendons are the two cords which can be felt at the back of the knee.

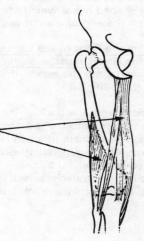

A third large group of muscles is sandwiched between the quadriceps and hamstrings. These muscles, the *adductors*, arise from the pubis. They are responsible for drawing the thighs together.

5.3 The lower leg and foot

The **tibia** is strong and weight-bearing. The **fibula** is slender being for muscle attachment only. The lower ends of these bones project on either side of the ankle where they enclose the **talus**.

The talus is a block of bone with a rounded upper surface which lies on the heel bone or *calcaneus*.

The foot turns up and down at the ankle between the talus and the tibia.

Movements which turn the sole of the foot in and out occur at the joint between the talus and the calcaneus.

The remaining five tarsal bones and the metatarsal bones are bonded strongly together, and to the talus and calcaneus, forming a concave bony arch.

This arched structure acts as a shock absorber. The arch is supported by deep muscles in the calf which insert via tendons into the sole of the foot and into the top of the arch.

Strong ligaments in the sole add further strength to the arch.

Downward movement of the foot, as in standing on tip-toe, is powered by muscles in the back of the calf, the **gastrocnemius** and the **soleus muscles**.

The gastrocnemius muscle arises from the lower femur. The soleus muscle arises from the tibia. They unite at the **Achilles tendon** which inserts into the calcaneus.

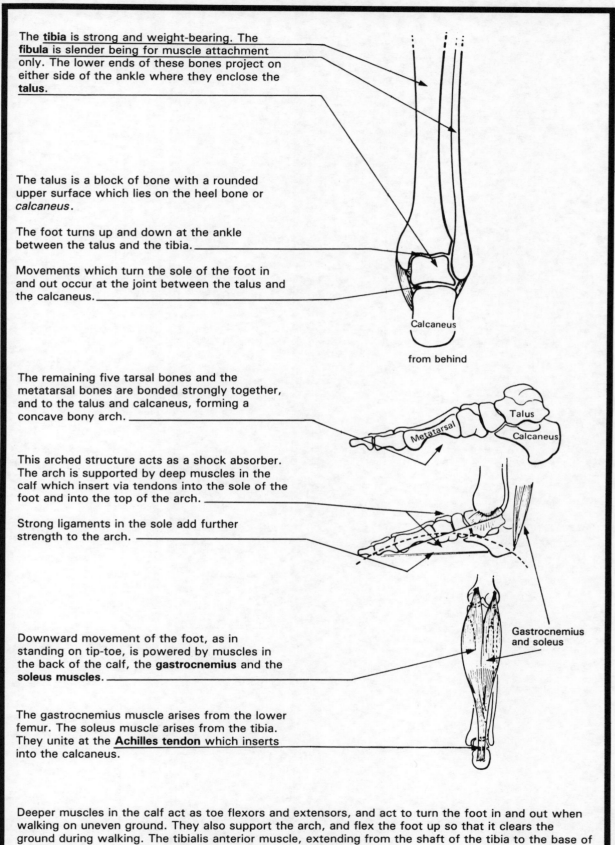

Calcaneus

from behind

Metatarsal

Talus

Calcaneus

Gastrocnemius and soleus

Deeper muscles in the calf act as toe flexors and extensors, and act to turn the foot in and out when walking on uneven ground. They also support the arch, and flex the foot up so that it clears the ground during walking. The tibialis anterior muscle, extending from the shaft of the tibia to the base of the first metatarsal, is also involved in movements of the foot.

TEST FIVE

1. (a) Name the three bones which are fused together to form the pelvic bone.

   _____

   (b) On which of these bones does one sit?

   _____

2. (a) What is the name of the joint at the front of the pelvis?

   _____

   (b) What type of joint is it?

   _____

3. Which of the statements on the right apply to the muscles listed on the left?

   (i)   The psoas and iliacus.   (a) Unlocks the knee.

   (ii) The gluteal muscles.   (b) Flex the knee.

   (iii) The quadriceps.   (c) Straightens the knee.

   (iv) The hamstrings.   (d) Help raise the legs alternately in walking.

   (v)   The popliteus.   (e) Flex the hip.

4. (a) What bones go to make up the knee joint?

   _____

   (b) What are the muscles, ligaments and tendons involved?

   _____

5. Arrange the following bones in their distal order (i.e. their distance from the body's centre).

   Tibia, calcaneus, femur, fibula, metatarsal bones, tarsal bones.

ANSWERS TO TEST FIVE

1. (a)   Ilium, ischium and pubis.

   (b)   Ischium.

2. (a)   Pubic symphysis.

   (b)   Cartilaginous.

3.

    (i)   The psoas and iliacus   (e)   flex the hip.

   (ii)   The gluteal muscles   (d)   help raise the legs alternately in walking.

  (iii)   The quadriceps   (c)   straighten the knee.

  (iv)   The hamstrings   (b)   flex the knee.

   (v)   The popliteus   (a)   unlocks the knee.

4. (a)   The knee joint is made up of femur, tibia and patella.

   (b)   The quadriceps wrap round the femur shaft, insert into the patella and then into the front of the tibia via the patellar tendon. Ligaments support the joint on either side, while inside its cavity crossing ligaments provide the joint with additional movement and stability.

5.   Femur, tibia, fibula, calcaneus, tarsal bones, metatarsal bones.

POST TEST

1. **What are the main characteristics of a synovial joint?**

   _____

   _____

2. **What is the name given to the centre of ossification at the end of a typical developing long bone?**

   _____

3. **Which of the descriptions on the right apply to the items listed on the left?**

   (i)   The periosteal collar.   (a)   Bone-removing cells.
   (ii) The lamellae.   (b)   Thin sheets of bone.
   (iii) The osteoclasts.   (c)   The first bone to appear on the cartilage model.

4. **Give two actions of the biceps muscle.**

   _____

5. **What are the names of the two proteins most intimately involved in muscle contraction?**

   _____

6. **What muscle patterns are represented by the diagrams below?**

A   B   C

   (i)   Pennate.

   (ii) Fusiform.

   (iii) Strap.

POST TEST

7. **Label four major skull bones.**

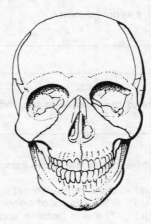

8. **Which of the major neck muscles fulfill functions other than that of flexing the neck?**

_____

_____

9. **Which of the descriptions on the right apply to the parts of the leg listed on the left?**

(i) The hamstrings.   (a) The muscles which flex the knee.

(ii) The menisci.   (b) Rounded projections at the end of the femur.

(iii) The condyles.   (c) Cartilage discs between the femur and tibia.

10(a) **Which two sets of muscles unite at the Achilles tendon?**

_____

(b) **Into which bone does the Achilles tendon insert?**

_____

ANSWERS TO POST TEST

1. (a) Most synovial joints allow a great deal of movement;

   (b) the articulating surfaces of bones are covered with smooth, hard cartilage;

   (c) the joint is enclosed in a fibrous capsule that is lined with synovial membrane;

   (d) the synovial membrane secretes lubricating synovial fluid into the joint cavity.

2. Epiphysis.

3. (i) The periosteal collar   (c)   is the first bone to appear on the cartilage model.

   (ii) The lamellae   (b)   are thin sheets of bone.

   (iii) The osteoclasts   (a)   are bone-removing cells.

4. Flexion (bending of elbow, lifting forearm).

   Supination (the winding, palm-up movement in using a screwdriver).

5. Actin and myosin.

6. A (iii) Strap.

   B (ii) Fusiform.

   C (i) Pennate.

ANSWERS TO POST TEST

7. Four of:

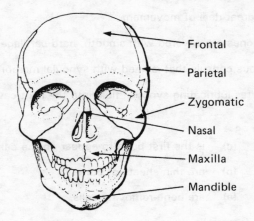

- Frontal
- Parietal
- Zygomatic
- Nasal
- Maxilla
- Mandible

8. The anterior strap muscles help form the floor of the mouth and support the larynx as well as flexing the neck.

9. (i) The hamstrings   (a)   are the muscles which flex the knee.

    (ii) The menisci   (c)   are cartilage discs between the femur and tibia.

    (iii) The condyles   (b)   are rounded projections at the end of the femur.

10. (a) The gastrocnemius muscle (arising from the lower femur) and the soleus (arising from the tibia) unite at the Achilles tendon.

    (b) The tendon inserts into the calcaneus.

Contents

The Special Senses

1. The chemical senses — smell and taste

1.1. Smell

The sense of smell helps greatly in detecting the flavour, as well as the aroma of suitable food. Its main function is probably protection.

Smell is a primitive sense; its primary connections are to those parts of the brain which were the earliest to evolve.

The sense of smell is relatively unimportant to man, but not to many animals.

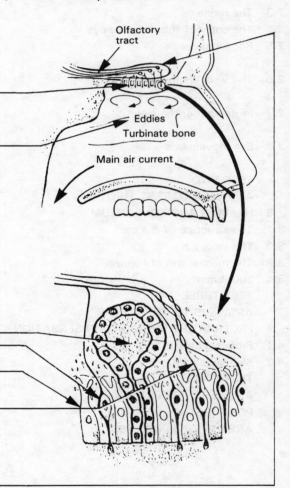

Olfactory tract

Eddies
Turbinate bone

Main air current

The sensory olfactory epithelium occupies a few square centimetres in the roof of the nose.

The current of inspired air does not pass directly over the sensory cells, but eddies of air reach them.

To be smelt substances must be volatile, and also soluble in lipids.

Molecules of smelling substances dissolve in the secretions of the local **mucous glands** and are detected by the sensory **olfactory** cells which have a fringe of blunt cilia and lie amid **supporting** cells.

Stimulation causes impulses to pass along the **nerve fibres** of the sensory cells.

These nerve fibres pierce the roof of the nose to enter the cranial cavity where they join the olfactory bulb.

From the **olfactory bulb** nerves in the olfactory tract pass to certain nuclei at the base of the brain, the *pyriform area*, and then by a complex path to an area of cerebral cortex in the cleft between the hemispheres, *the cingulate gyrus*.

1.2. Taste

Taste is a much cruder sense than smell. Most of the sensation we call taste is due to the smell and feel of food within the mouth.

The receptors for taste are the taste buds which are found mainly around the edge of the upper surface of the tongue, and also on the soft palate.

Taste buds consist of a collection of stave-like cells opening by a tiny pore on the surface of small projections or **papillae**.

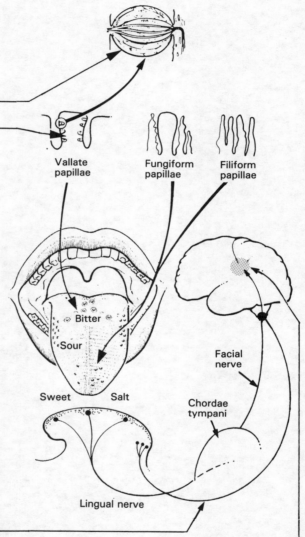

Vallate papillae

Fungiform papillae

Filiform papillae

Bitter

Sour

Sweet   Salt

Facial nerve

Chordae tympani

Lingual nerve

The papillae provide the tongue with its rough surface.

Three types of papillae are found on the tongue, but the taste buds in each have the same structure.

Different areas of the tongue tend to detect different types of taste.

Four basic tastes are usually described — bitter, sweet, sour and salt. However some people cannot distinguish between bitter and sour, and some can identify different types of sweetness. There is no obvious connection between chemical structure and taste.

Impulses from the taste buds on the *front* part of the tongue travel in fibres which pass through three different nerves on the way to the brain stem.

Impulses from the taste buds on the *back* of the tongue travel in fibres in the **glossopharyngeal nerve** to the brain stem.

Further nerve fibres carry the sensation of taste from the brain stem to the parietal lobe.

TEST ONE

1. What is involved in taste?

_____

_____

2. Which of the statements on the right apply to the items on the left?

(a) Fungiform papillae.

(b) The glossopharyngeal nerve.

(c) Taste buds.

(d) The lingual nerve.

(i) Collection of stave-like cells.

(ii) Projections on the surface of the tongue.

(iii) Carries impulses from the front of the tongue to the brain stem.

(iv) Carries impulses from the back of the tongue to the brain stem.

3. Which parts of the tongue respond to which of the four basic tastes?

	Bitter	Sweet	Sour	Salt
(a) The back of the tongue	( )	( )	( )	( )
(b) The sides of the tongue	( )	( )	( )	( )
(c) The front of the tongue	( )	( )	( )	( )

ANSWERS TO TEST ONE

1. Taste is compounded of the response of the taste buds together with response to the feel of an ingested substance (its roughness, smoothness, heat, cold) and its smell.

2. (a) Fungiform papillae   (ii) are projections on the surface of the tongue.

     (b) The glossopharyngeal nerve   (iv) carries impulses from the back of the tongue to the brain stem.

     (c) Taste buds   (i) are a collection of stave-lke cells.

     (d) The lingual nerve   (iii) carries impulses from the front of the tongue to the brain stem.

3.

	Bitter	Sweet	Sour	Salt
(a) The back of the tongue	(√)	( )	( )	( )
(b) The sides of the tongue	( )	( )	(√)	( )
(c) The front of the tongue	( )	(√)	( )	(√)

2. Vision

2.1. The structure of the eye

The eye is a fluid filled sphere about 24 mm in diameter. Its basic structure is that of a camera, with a focusing system, a mechanism to control light entry, a light sensitive layer, and a dark internal coating to limit the scatter of light.

The eye consists of three layers:

> an inner light sensitive nervous layer, the **retina**,
> a middle vascular and pigmented **choroid layer**,
> a thick outer coat, the **sclera**.

Light enters the eye by a transparent window in the sclera, the **cornea**.

It passes through the **pupil** which is a hole in a muscular screen called the **iris**. The iris is pigmented. It can contract and dilate to vary the amount of light entering the eye.

The light is focused by an elastic **lens**, just behind the iris. Ligaments secure the lens to the **ciliary body**, the anterior part of the choroid layer.

The light is focused as an inverted image on the **retina** at the back of the eye. The retina is almost transparent and the interior of the eye has a dark brown sooty appearance due to the pigmented and vascular choroid layer.

Nerve fibres from the retina leave, through a group of perforations in the sclera, at the **optic disc** to form the **optic nerve** to the brain.

The space between the lens and the cornea, the *anterior chamber* of the eye, is filled with watery **aqueous humour**. The aqueous humour is secreted by the ciliary body under pressure. The pressure is one of the factors maintaining the shape of the globe.

The **canal of Schlemm** encircles the junction of the cornea and sclera. It lies deep in the sclera at the edge of the anterior chamber of the eye.

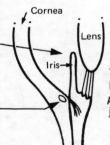

Cornea

Lens

Iris

The main cavity of the eye between the lens and retina, the *posterior chamber*, is filled with jelly-like *vitreous humour*.

Aqueous humour drains into the canal of Schlemm, and from there into the veins of the sclera. Obstruction of this drainage is one of the causes of *glaucoma*, abnormally high pressure within the globe.

2.2. Focusing mechanism

Most of the focusing power of the eye is due to refraction of light by the cornea. Refraction of light by the lens of the eye is of great importance; the curvature of the lens can vary so that light is always focused on to the retina.

The lens is transparent and pale yellow. It is kept flattened by the normal tension of the eyeball, maintained by the **suspensory ligaments.**

The shape of the lens is altered by the **ciliary muscle**, which is contained in the ciliary body.

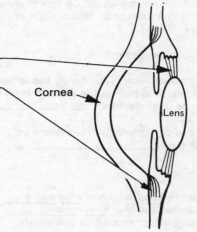

When it contracts the ciliary muscle draws the ciliary body forwards, relaxing the tension on the lens and allowing it to bulge. Light from near objects can then be focused on to the retina.

The ciliary muscle relaxes when the eye has to focus light from distant objects on to the retina.

The ciliary muscle is supplied by parasympathetic nerve fibres of the oculomotor nerve.

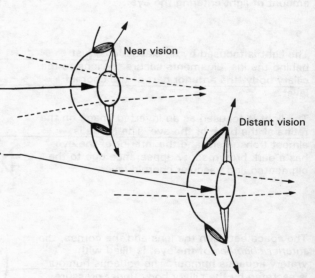

The **iris** is a screen of smooth muscle perforated by the **pupil.**

The size of the pupil changes with changing light conditions, dilating in the dark and constricting in bright light so that overstimulation of the retina is prevented. The size of the pupil is governed by the contraction of radial dilator and circular constrictor muscle fibres in the iris. These fibres are supplied by parasympathetic nerves from the third cranial nerve.

2.3. The retina

The retina lines the back half of the interior of the eye.

The retina is nourished by blood vessels which form a tree-like pattern on its surface.

There are no blood vessels over the **macula lutea** which lies at the middle of the back of the eye.

At the centre of the macula lutea, directly opposite the lens, is a shallow pit, the *fovea* where some of the light sensitive cells are crowded together.

On the inner (nasal) side of the macula lutea is the **optic disc** where nerve fibres leave the eye. There are no light sensitive cells at this point, which accounts for the blind spot 15° to the outside of the central vision. One is normally unaware of the blind spot.

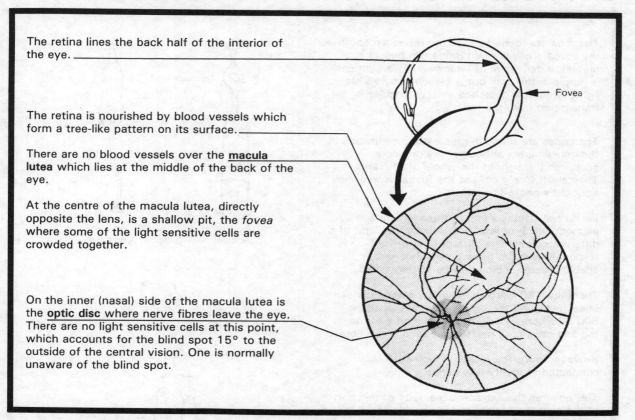

Fovea

2.4. The entry of light into the eye

Light entering the eye first passes through the **neurones** on the retinal surface, before stimulating the light-sensitive cells.

The light-sensitive cells are of two kinds:

the rods, which can function in dim light, but which do not register colour;

the cones, which are responsible for colour vision.

Impulses from rods and cones are transmitted first to **bipolar cells**, and then to **ganglion cells**, whose axons cross the retina and leave the eye at the optic disc as the optic nerve.

Pigment granules in cells in the choroid layer can move up and down the processes between the rods and cones to limit the scatter of light from one cell to the next.

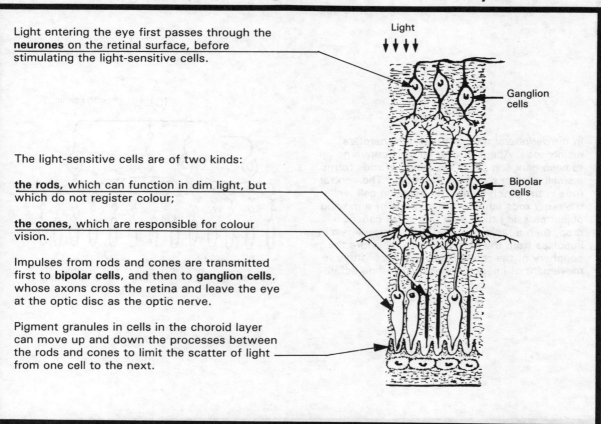

Light

Ganglion cells

Bipolar cells

2.5. Rods and cones

The rods are found all over the retina except in the fovea. They contain rhodopsin (visual purple), a derivative of vitamin A. This pigment is purple in the dark but is bleached by visible light. When it is bleached a nerve impulse is transmitted from the rod.

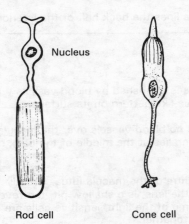

Nucleus

Rod cell   Cone cell

The cones are found in greatest concentration in the macula lutea although they do occur scattered throughout the whole of the retina. They are the only cells at the fovea, where they are tightly packed together.

Each cone contains one of three separate pigments each of which is bleached by light of a different colour. The different patterns of response from the three types of cone cell enable different colours to be distinguished.

There are 120 million rods and 7 million cones in each eye. As the optic nerve only contains 800 000 fibres, considerable numbers of the rod and cone cells must share each nerve fibre.

However in the *fovea* each bipolar cell is connected to a very few cones.

This ensures the extreme sensitivity of this part of the eye; the normal eye is capable of resolving very fine detail.

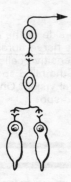

In the *peripheral* part of the retina there are mainly rods. About 300 or so rods converge on to each ganglion cell. This family of rods forms a small circular patch on the retina. The central rods in the patch excite the ganglion cell while the outer rods inhibit it. The image of a moving object crossing the patch first of all causes a drop, then a rapid rise, and then a drop again in impulses from the ganglion cell. Therefore the periphery of the retina is especially sensitive to movement of images rather than to fine detail.

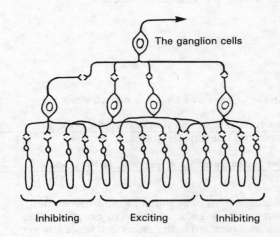

The ganglion cells

Inhibiting   Exciting   Inhibiting

2.6. The visual pathways

40% of the sensory input to the brain is visual.

The areas seen by each eye overlap considerably, but they are slightly different. The temporal visual field of each eye is larger than its nasal visual field because of the nose and cheeks.

The rays of light from objects in the temporal visual field fall on the nasal side of the retina, and those from objects in the nasal visual field fall on the temporal side of the retina.

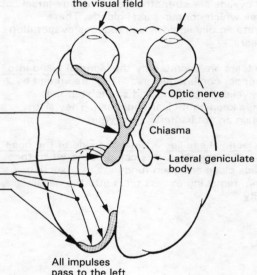

Temporal visual field | Nasal visual field

Nasal visual field | Temporal visual field

Fixation point

Left eye   Right eye

From the right half of the visual field

The nerve fibres from the nasal sides of the two retinas cross over at the union of the optic nerves, the **optic chiasma**, which is inside the cranial cavity. The nerve fibres from the temporal sides of the retinas do not cross.

All the fibres synapse at the **lateral geniculate** bodies and then pass by broad **optic radiations** through the brain to the **visual cortex** at the top of the occipital lobes.

Optic nerve

Chiasma

Lateral geniculate body

All impulses pass to the left cortex

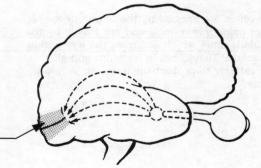

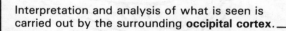

Interpretation and analysis of what is seen is carried out by the surrounding **occipital cortex**.

The entire visual world is thus divided by a central vertical line. All impulses originating from the left visual field are 'seen' by the right half of the brain and vice versa.

2.7. Protection of the eye

The eye is contained within a bony **orbit**. Twice the diameter of the eye, it is padded with fat and contains the **muscles** which move the eye, and the gland which secretes tears.

The space between the eyelids and eyeball is the **conjunctival** sac.

It is lined by the conjunctiva, a squamous epithelium covering the interior of the eyelids and the surface of the eyeball.

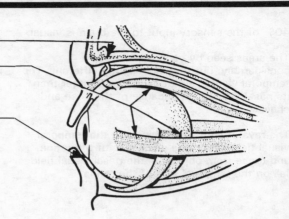

The eyelids are strengthened by fibrous **tarsal plates** which contain tarsal glands. These secrete an oily liquid which reduces evaporation of tears.

The tears are secreted by the **lacrimal gland** into the upper conjunctival sac. They are spread by the eyelids in blinking and so moisten the cornea keeping it clear and clean. They also contain an antibacterial lysozyme.

The **lacrimal sac** lies against the side of the nose just inside the orbit. It is compressed when the eyelids close and then functions as a suction pump, removing excess tears into the nasal cavity.

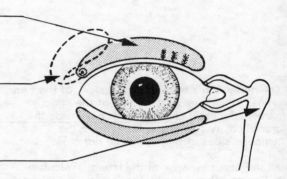

The eyelids are opened by the action of the **levator palpebrae muscle** and are closed by the **orbicularis oculi** which encircles the eye within the eyelids. They close in blinking, and also close reflexly to protect the eye from foreign objects.

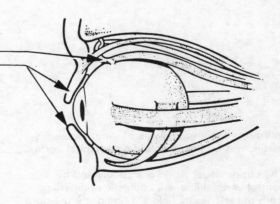

2.8. Eye movements

The eyeball is rotated in different directions by six muscles.

Four recti muscles arise at the back of the orbit on the nasal side, at the exit of the optic nerve.

The **medial rectus** turns the eye inwards.

The **lateral rectus** turns the eye outwards.

The **superior rectus** rotates the eye upwards and inwards.

The **inferior rectus** rotates the eye downwards and inwards.

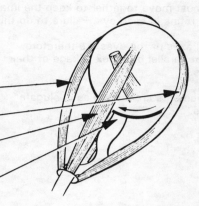

Right eye
viewed from above

The other two muscles are the superior and inferior oblique muscles.

The **superior oblique** passes through a pulley. Its tendon is inserted far back on the eye. It turns the eye downwards and outwards.

The **inferior oblique** turns the eye upwards and outward. It is the only muscle which arises from the front of the orbit.

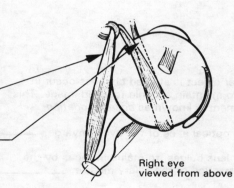

Right eye
viewed from above

Most of the eye muscles are supplied by the third (oculomotor) cranial nerve. The lateral rectus is supplied by the sixth cranial nerve and the superior oblique by the fourth cranial nerve.

The eye muscles and eye movements are summarised by the diagram below.

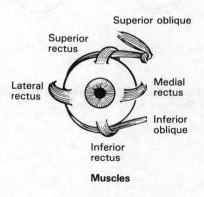

Muscles

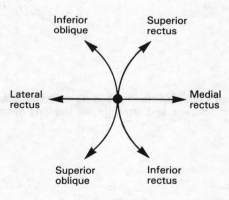

Movements

55

The eyes must move together to keep the image of the object under observation on the same relative part of the retina in each eye. Failure to do this results in double vision.

The optical axes of the eyes are therefore usually kept parallel by reflex linkage of their movements.

These movements are known as conjugate movements.

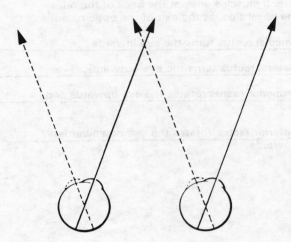

If a near object is studied changes occur in order to maintain a single focused image. This adjustment is known as *accommodation*

1. The optical axes of the eye converge.

2. The lens bulges, to maintain focus by the contraction of the ciliary muscle.

3. The pupil constricts in bright light, (admitting less light) and dilates in dim light (admitting more light).

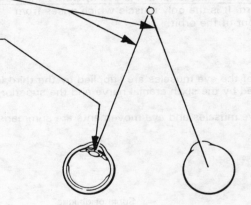

Children can keep objects in focus down to a distance of 7 cm. This distance increases to about 40 cm in middle age and later life, as the lens becomes less elastic.

TEST TWO

1. Tick to show the characteristics of the different layers of the wall of the eyeball:

	Inner coat	Middle coat	Outer coat
Is made of fibrous tissue.	( )	( )	( )
Is vascular.	( )	( )	( )
Contains the cornea.	( )	( )	( )
Contains the light-sensitive cells.	( )	( )	( )
Includes the iris.	( )	( )	( )
Is pigmented.	( )	( )	( )

2. Which of the statements on the right apply to the parts of the eye listed on the left?

   (i) The aqueous humour.   (a) Lies against the retina.
   (ii) The vitreous humour.   (b) Produced by the ciliary body.
   (iii) The suspensory ligaments. (c) Support the lens.

3. Tears help to protect the eyes; what helps to ensure the efficiency of this protection?

   _____

   _____

   _____

4. Which of the statements on the right apply to the parts of the retina listed on the left?

   (i) The optic disc.   (a) Is free of rods.
   (ii) The macula lutea.   (b) Is free of nerve cells.
   (iii) The fovea.   (c) Is free of blood vessels.

5. Tick the appropriate brackets to show to which cells the following statements best apply.

	Rods	Cones
Contain rhodopsin (visual purple).	( )	( )
There are seven million in each eye.	( )	( )
Found mainly in the periphery of the retina.	( )	( )
Function well in dim light.	( )	( )
Contain three types of visual pigments.	( )	( )

6. Which of the statements on the right apply to the eye muscles listed on the left?

   (i) The lateral rectus.   (a) Moves the eye inwards and downwards.
   (ii) The superior oblique.   (b) Moves the eye outwards.
   (iii) The inferior rectus.   (c) Moves the eye upwards and outwards.
   (iv) The inferior oblique.   (d) Moves the eye downwards and outwards.

7. With reference to the eyes, what is meant by:

   (a) conjugate movements
   (b) accommodation.

   _____

ANSWERS TO TEST TWO

1.

	Inner coat	Middle coat	Outer coat
Is made of fibrous tissue.	( )	( )	(√)
Is vascular.	( )	(√)	( )
Contains the cornea.	( )	( )	(√)
Contains the light-sensitive cells.	(√)	( )	( )
Includes the iris.	( )	(√)	( )
Is pigmented.	( )	(√)	( )

2.
  (i) The aqueous humour   (b) is produced by the ciliary body.
  (ii) The vitreous humour   (a) lies against the retina.
  (iii) The suspensory ligaments   (c) support the lens.

3. The tarsal glands secrete an oily liquid which reduces tear evaporation, and when the eyelids are closed thereby lessening the need for tears, the lachrymal sac removes excess tears into the nasal cavity.

4.
  (i) The optic disc   (b) is free of nerve cells.
  (ii) The macula lutea   (c) is free of blood vessels.
  (iii) The fovea   (a) is free of rods.

5.

	Rods	Cones
Contain rhodopsin (visual purple).	(√)	( )
There are seven million in each eye.	( )	(√)
Found mainly in the periphery of the retina.	(√)	( )
Function well in dim light.	(√)	( )
Contain three types of visual pigments.	( )	(√)

6.
  (i) The lateral rectus   (b) moves the eye outwards.
  (ii) The superior oblique   (d) moves the eye downwards and outwards.
  (iii) The inferior rectus   (a) moves the eye inwards and downwards.
  (iv) The inferior oblique   (c) moves the eye upwards and outwards.

7. (a) Conjugate movements refer to the reflexes by which the image of an observed object is kept on the same relative part of each retina.
  (b) Accommodation is the adjustment in vision required in keeping a near image focused.

3. Hearing and equilibrium

3.1. The structure of the ear

The ear is made up of three parts:

(1) the external ear, which consists of the **pinna**, (the visible ear), and the **external auditory meatus**, which is about 25 mm long and ends at the ear drum or **tympanic membrane**;

(2) the middle ear which is an air filled cavity containing the three auditory ossicles **the malleus**, **the incus**, **the stapes**;

(3) the inner ear which contains the **cochlea**, the sense organ for hearing, the **semicircular canals**, which detect rotation of the head, and the **vestibule**, which is concerned with position.

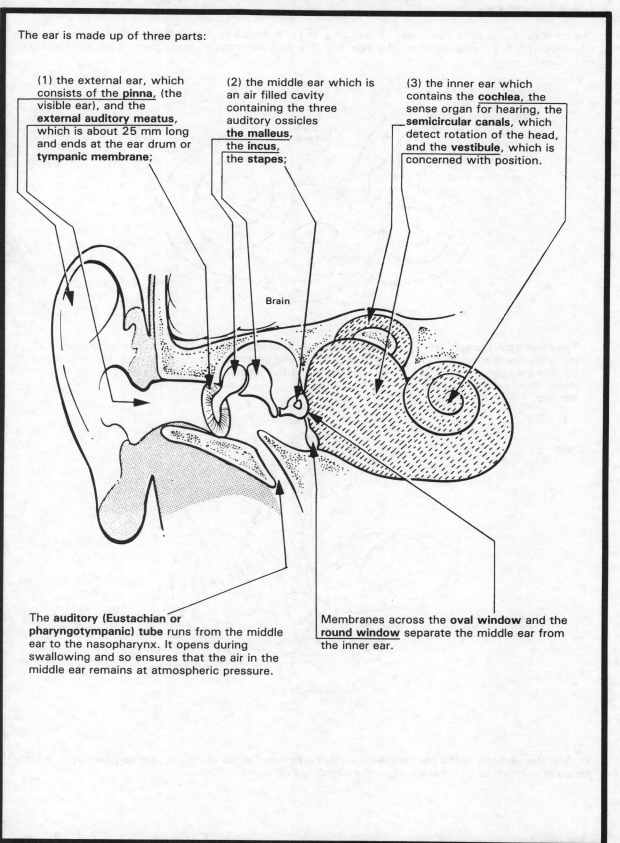

Brain

The **auditory (Eustachian or pharyngotympanic) tube** runs from the middle ear to the nasopharynx. It opens during swallowing and so ensures that the air in the middle ear remains at atmospheric pressure.

Membranes across the **oval window** and the **round window** separate the middle ear from the inner ear.

3.2. The inner ear

The organs of hearing and balance are intimately associated in the inner ear. They share the same cranial nerve (the eighth).

The inner ear consists of a delicate, complicated, fluid-filled tube known as the membraneous labyrinth. This lies inside the bony labyrinth, which is a chamber buried in the *petrous bone* in the floor of the skull.

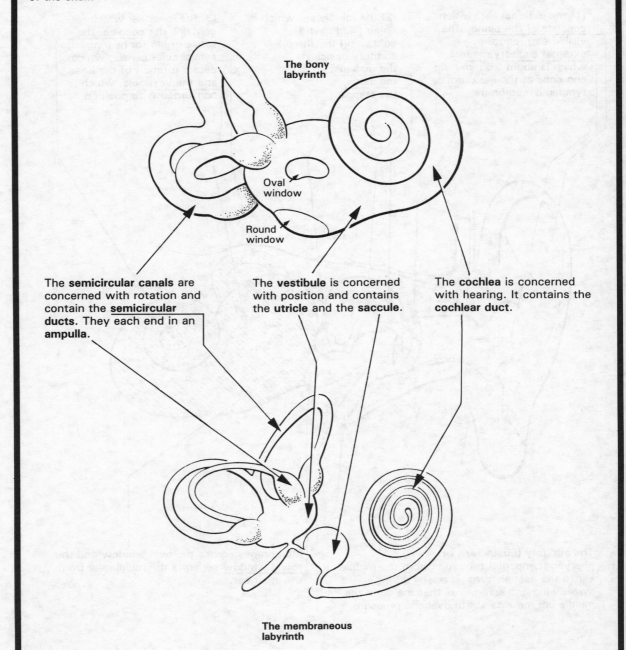

The bony labyrinth

Oval window

Round window

The **semicircular canals** are concerned with rotation and contain the semicircular ducts. They each end in an **ampulla**.

The **vestibule** is concerned with position and contains the **utricle** and the **saccule**.

The **cochlea** is concerned with hearing. It contains the **cochlear duct**.

The membraneous labyrinth

Each of the sections of the membraneous labyrinth contains areas of delicate sensory hair cells, which send nerve impulses to the brain along the eighth cranial nerve.

Despite their complexity all the sensory areas of the inner ear have the same basic plan.

Inside the bony labyrinth there is a clear watery fluid rich in sodium ions — **the perilymph. The membraneous labyrinth is a thin tube which floats in the perilymph. It contains a fluid rich in potassium ions — the endolymph.**

In some areas the membraneous labyrinth is thickened by a patch of sensory epithelium composed of thousands of **hair cells.**

These sensory areas are covered by a firm **gelatinous mass**. Relative movement between the hair cells and the gelatinous mass causes changes in the rate at which nerve impulses are discharged.

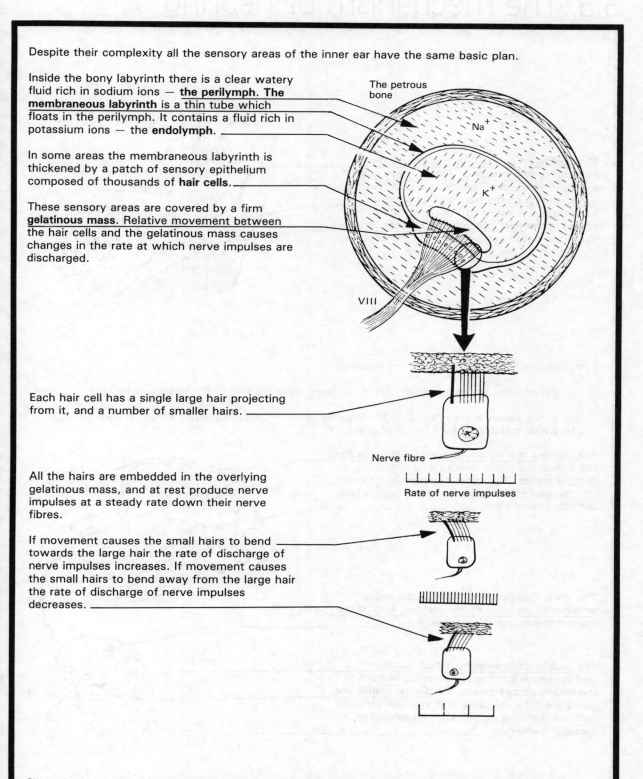

The petrous bone

Na^+

K^+

VIII

Each hair cell has a single large hair projecting from it, and a number of smaller hairs.

Nerve fibre

Rate of nerve impulses

All the hairs are embedded in the overlying gelatinous mass, and at rest produce nerve impulses at a steady rate down their nerve fibres.

If movement causes the small hairs to bend towards the large hair the rate of discharge of nerve impulses increases. If movement causes the small hairs to bend away from the large hair the rate of discharge of nerve impulses decreases.

Since the hair cells point in different directions in different parts of the sensory epithelium the nerve impulses provide the brain with accurate information on the direction and amount of displacement of the gelatinous body.

Modifications of this basic plan occur in the cochlea (which detects sound vibrations), in the utricle and saccule (which detect position), and in the semicircular canals (which detect rotation).

3.3. The mechanism of hearing

The **auricle** funnels sound waves into the **external auditory meatus**, where they cause the ear drum to vibrate. The malleus is bound to the **ear drum** and so the ossicles vibrate and transmit the vibrations to the footplate of the stapes in the **oval window** of the inner ear. The vibrations are then transmitted to the cochlea.

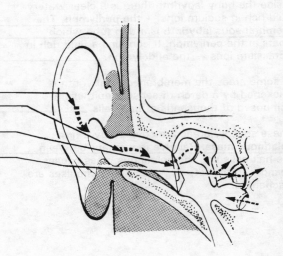

Amplification occurs in the middle ear because:

(i) the ossicles act as a system of levers and increase the size of the vibrations by half;

(ii) all the sound energy falling on the large ear drum is concentrated on the small oval window which amplifies it by a factor of about 15.

The **cochlea** is a coiled tube with a bulbous base like a snail shell. It is divided into three compartments. The upper compartment joins into the rest of the labyrinth at the base where the oval window lies.

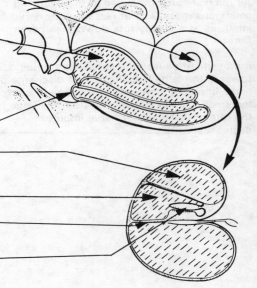

The lower compartment ends at the **round window** at the base of the cochlea. Both these compartments contain **perilymph**.

The **middle compartment**, which contains endolymph, is completely sealed. The floor of the middle compartment, the **basilar membrane**, bears the organ of Corti, a strip of sensory hair cells whose hairs extend into the overlying **tectorial membrane**.

Vibrations are transmitted by the footplate of the stapes to the upper compartment of the cochlea. From there the vibrations are transmitted to the basilar membrane and from there through the lower compartment to the round window.

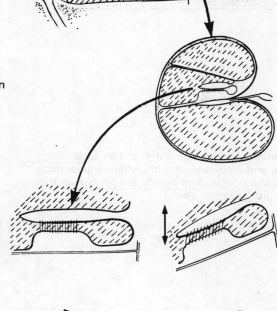

The round window vibrates in the opposite direction to the oval window and so the fluids in the inner ear can vibrate freely.

Up-and-down vibrations of the basilar membrane bend the hairs in the cells of the organ of Corti. This causes the generation of nerve impulses to increase. The inner and outer hair cells differ in their sensitivity to movement. This difference forms the basis of loudness discrimination.

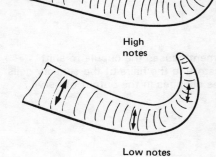

The pattern of vibration in the basilar membrane varies with the pitch of the sound enabling pitch to be discriminated. High pitched sounds cause vibration of the basal part of the membrane. Low-pitched sounds cause all of the membrane to vibrate.

High notes

Low notes

Nerve fibres from the organ of Corti have their cell bodies in the spiral ganglion close to the cochlea. Nerve impulses pass via the eighth cranial nerve and the brain stem to the temporal lobe where sounds are analysed.

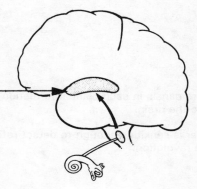

3.4. Equilibrium

Rotation of the head is detected by three **semicircular ducts**. These are curved tubes which open into the **utricle**. They lie at right angles to each other, like three sides at a corner of a box. Each semicircular duct widens into an <u>ampulla</u>, at one end.

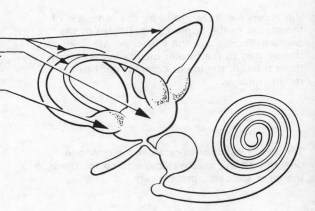

The ampulla contains a patch of <u>**hair cells, the crista**</u>, and an overlying dome-shaped gelatinous mass, the cupula. <u>**The cupula** almost blocks the duct, but is free to swing.</u>

When rotation occurs in the plane of a canal the endolymph fluid in the canal tends to lag behind due to inertia.

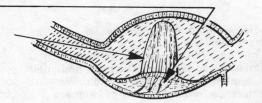

Nerve impulses

No movement

The movement causes the cupula to shift slightly, distorting the hairs of the sensory cells. This causes changes in the output of nerve impulses.

Rotation to left

Rotation to right

Since there are canals in each plane information on rotation of the head in any plane will be transmitted to the brain.

The human ear is sensitive enough to detect rotation which is so slow that it takes one minute to complete one revolution.

Movement in a straight line, and position, are detected by the utricle and saccule. In each there is a patch of cells, the macula. The macula of the utricle and the **macula** of the saccule lie at right angles to each other.

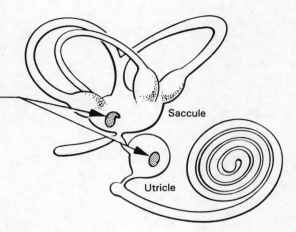

Saccule

Utricle

A **gelatinous otolith membrane** which contains heavy calcium salt crystals, or **otoliths**, overlays each macula.

When the position of the head is changed the heavy otoliths move.

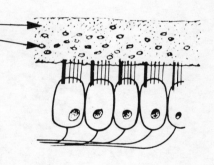

All large hairs on left

Direction of maximum stimulation

The hair cells in each macula are so laid out that the precise direction and degree of movements of the otoliths can be detected and transmitted to the brain.

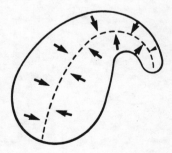

Impulses from the semicircular canals, utricle and saccule are carried via the eighth cranial nerve to the brain stem. Impulses are relayed from here to the muscles in the legs, trunk and neck, and to the eye muscles, so that the body can immediately compensate for changes in its position. Some impulses are relayed to the temporal lobe cortex where they reach consciousness.

Motion due to a straight-line, arm-position... is
detected by the utricle and saccule. In each
there is a patch of cells, the macula. The
macula of the utricle, and the macula of the
saccule, is at right angles to each other.

Saccule

Duct

A gelatinous otolith membrane, which contains
heavy calcium salt crystals, of otoliths, overlays
each macula.

When the position of the head is changed, the
heavy otolith move...

All the hair of...

Direction of maximum stimulation

The hair cells in each macula line are so laid out that
the precise direction and degree of movement
of the otoliths can be detected and transmitted
to the brain.

impulses continue without cease, until... in the eighth cranial nerve to
the brain. Impulses are relayed from there to the muscles in the legs, trunk and neck, and to the
eye muscles so that the body can immediately compensate for changes in its position. Some impulses
are relayed to the cerebral cortex, where they reach consciousness.

66

TEST THREE

ANSWERS TO TEST THREE

1. Indicate the ways in which sound is transmitted through the three parts of the ear by placing ticks in the appropriate brackets.

	External ear	Middle ear	Inner ear
(a) By vibrations in air.	( )	( )	( )
(b) By vibrations in fluid.	( )	( )	( )
(c) Mechanically.	( )	( )	( )

2. Which of the statements on the right apply to the parts of the ear listed on the left?

(i) The auditory tube.	(a) Is concerned with hearing.
(ii) The cochlea.	(b) Is filled with endolymph.
(iii) The vestibule.	(c) Contains the utricle.
(iv) The membraneous labyrinth.	(d) Maintains atmospheric pressure in the middle ear.

3. Indicate the locations of the structures responsible for the properties below by placing ticks in the appropriate brackets.

	Semicircular canals	Cochlea	Middle ear	Vestibule
(a) Amplification of sound waves.	( )	( )	( )	( )
(b) Sensing of static position.	( )	( )	( )	( )
(c) Perception of sound waves.	( )	( )	( )	( )
(d) Sensing rotation.	( )	( )	( )	( )

4. Complete the following:

The perilymph is a fluid rich in _____ ions and found inside the _____ of the ear.
The endolymph is a fluid rich in _____ ions and contained within the _____
which floats in _____ .

5. Indicate where the following structures are found by placing ticks in the appropriate brackets.

	Semicircular canals	Cochlea	Utricle
(a) Tectorial membrane.	( )	( )	( )
(b) Ampulla.	( )	( )	( )
(c) Organ of Corti.	( )	( )	( )
(d) Otoliths.	( )	( )	( )
(e) Cupula.	( )	( )	( )
(f) Basilar membrane.	( )	( )	( )

ANSWERS TO TEST THREE

1.

	External ear	Middle ear	Inner ear
(a) By vibrations in air.	(√)	( )	( )
(b) By vibrations in fluid.	( )	( )	(√)
(c) Mechanically.	( )	(√)	( )

2.
(i) The auditory tube   (d) maintains atmospheric pressure in the middle ear.
(ii) The cochlea   (a) is concerned with hearing.
(iii) The vestibule   (c) contains the utricle.
(iv) The membraneous labyrinth   (b) is filled with endolymph.

3.

	Semicircular canals	Cochlea	Middle ear	Vestibule
(a) Amplification of sound waves.	( )	( )	(√)	( )
(b) Sensing of static position.	( )	( )	( )	(√)
(c) Perception of sound waves.	( )	(√)	( )	( )
(d) Sensing rotation.	(√)	( )	( )	( )

4. The perilymph is a fluid rich in *sodium* ions and found inside the *bony labyrinth* of the ear. The endolymph is a fluid rich in potassium ions and contained within the *membraneous labyrinth* which floats in *perilymph.*

5.

	Semicircular canals	Cochlea	Utricle
(a) Tectorial membrane.	( )	(√)	( )
(b) Ampulla.	(√)	( )	( )
(c) Organ of Corti.	( )	(√)	( )
(d) Otoliths.	( )	( )	(√)
(e) Cupula.	(√)	( )	( )
(f) Basilar membrane.	( )	(√)	( )

4. Proprioceptors and skin sensation

4.1. Proprioceptors

Proprioceptors are sensory receptors which detect tension in muscles, joints, ligaments and tendons. They inform the brain of the positions of the various parts of the body, and so enable the movements of the limbs to be accurately controlled.

The proprioceptors include the following:

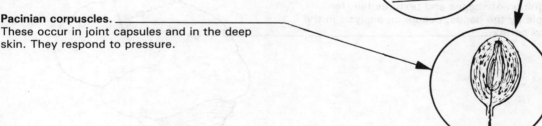

Muscle spindles ————————————
These are about 1–2 mm long and lie between the fibres of skeletal muscle. They fire off nerve impulses when stretched, and so provide information on muscle tension. They themselves contain a few tiny striated muscle fibres with their own independent nerve supply. When these fibres contract they cause the sensitivity of the muscle spindles to vary.

Golgi tendon organs. ————————————
These are stimulated by alteration of the tension in tendons.

Pacinian corpuscles. ————————————
These occur in joint capsules and in the deep skin. They respond to pressure.

All of these receptors respond to a change in conditions by causing the rate at which nerve impulses are discharged to change. This changed rate rapidly fades away under constant conditions as the receptor adapts to the stimulus. They are therefore much more sensitive to changes than to steady state conditions.

Information on position is transmitted to the cerebellum where it is used to maintain posture by causing unconscious changes to the muscles of the limbs and trunk.

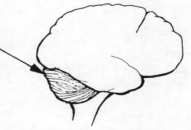

4.2. Skin sensation

The skin is sensitive to a variety of stimuli. Its sensitivity varies; the fingertips and face are very sensitive, the skin of the back is least sensitive.

In hairless skin touch is registered by **Merkel's discs** ⎯⎯⎯⎯⎯⎯⎯⎯⎯⎯⎯⎯

and **Meissner's corpuscles.** ⎯⎯⎯⎯⎯⎯⎯⎯

In hairy skin there are sets of nerve fibres around hair roots which respond to movement of the hairs. ⎯⎯⎯⎯⎯⎯⎯⎯⎯⎯⎯⎯

There are numerous free nerve endings in skin and deeper tissues, which are responsible for the appreciation of touch, pain and temperature. ⎯⎯

The deep dermis contains **Pacinian corpuscles.** ⎯⎯

Sensations from the body surface reach consciousness in the **post central area** of the cortex.

Recognition of shapes and textures felt, for example by the hands, occurs on analysis in the **parietal cortex.** ⎯⎯⎯⎯⎯⎯⎯⎯⎯⎯⎯⎯

In the viscera (gut, lung, etc.) pain is caused by distension rather than by mechanical stimuli. It is not well localised, but felt diffusely.

TEST FOUR

1. To what stimuli do the following proprioceptors respond?

	Muscle spindle	Golgi organ	Pacinian corpuscle
(a) Change in length of skeletal muscle.	( )	( )	( )
(b) Change in length of tendon.	( )	( )	( )
(c) Pressure.	( )	( )	( )

2. Indicate which of the names in the list below refer to the features labelled on the diagram alongside, by placing the appropriate letter in the brackets.

1. Pacinian corpuscle.   ( )

2. Free nerve ending.   ( )

3. Nerve fibre detecting
   hair movement.   ( )

4. Meissner's corpuscle.   ( )

5. Merkel's disc.   ( )

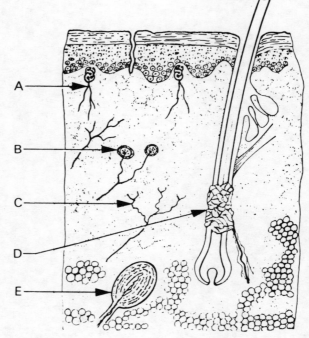

ANSWERS TO TEST FOUR

1.

	Muscle spindle	Golgi organ	Pacinian corpuscle
(a) Change in length of skeletal muscle.	(√)	( )	( )
(b) Change in length of tendon.	( )	(√)	( )
(c) Pressure.	( )	( )	(√)

2. 1. Pacinian corpuscle. (E)

2. Free nerve ending (C)

3. Nerve fibre detecting hair movement. (D)

4. Meissner's corpuscle. (A)

5. Merkel's disc. (B)

POST TEST

1. **Indicate which of the names in the list below refer to the features labelled on the diagram by placing the appropriate letters in the brackets.**

    1. Olfactory cells.   ( )

    2. Mucous gland.   ( )

    3. Olfactory epithelium.   ( )

    4. Olfactory tract.   ( )

    5. Olfactory bulb.   ( )

    6. Turbinate bones.   ( )

2. (a) What is the name of the sensory receptor for taste? _____

    (b) Which taste is best perceived at the tip of the tongue? _____

    (c) Name one of the nerves carrying taste fibres. _____

3. **Indicate which of the names in the list below refer to the features labelled on the diagram, by placing the appropriate letters in the brackets.**

    1. Vitreous humour.   ( )

    2. Optic nerve.   ( )

    3. Aqueous humour.   ( )

    4. Cornea.   ( )

    5. Iris.   ( )

    6. Sclera.   ( )

    7. Ciliary body.   ( )

    8. Suspensory ligament.   ( )

    9. Choroid.   ( )

    10. Lens.   ( )

4. **Label the muscles which produce the movements shown.**

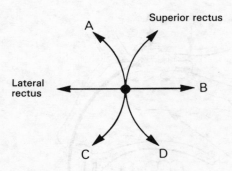

5. (a) **What is meant by a 'humour' in relation to the eye?**

_____

(b) **The eye contains two such 'humours', what are their names and where are they found?**

_____

_____

6. **Convergence of the eye's optical axes, bulging of the lens, constriction of the pupil; when, and for what purpose, do these processes take place?**

_____

_____

7. (a) **What is the name of the structure illustrated alongside?**

(b) **Indicate which of the names in the list below refer to the features labelled on the diagram, by placing the appropriate letters in the brackets.**

1. Saccule.   ( )

2. Utricle.   ( )

3. Cochlear duct.   ( )

4. Ampulla.   ( )

5. Semicircular duct.   ( )

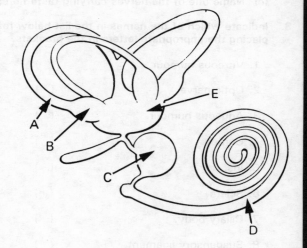

POST TEST

8. **Arrange the following organs in the order in which sound waves reach them.**

    (a) The oval window.
    (b) The ear drum.
    (c) The cochlea.
    (d) The auricle.
    (e) The external auditory meatus.

9. **Indicate which of the structures listed below are involved in the detection of rotation, position and sound by placing ticks in the appropriate brackets.**

	Rotation	Position	Sound
(a) Tectorial membrane.	( )	( )	( )
(b) Macula of utricle.	( )	( )	( )
(c) Semicircular ducts.	( )	( )	( )
(d) Otoliths.	( )	( )	( )
(e) Basilar membrane.	( )	( )	( )

10. **Indicate the names and locations of the proprioceptors shown below by placing the appropriate letters in the brackets.**

    1. Pacinian corpuscle. ( )

    2. Golgi organ. ( )

    3. Muscle spindle. ( )

    4. In skeletal muscle. ( )

    5. In tendons. ( )

    6. Around joints. ( )

    A

    B

    C

ANSWERS TO POST TEST

1.
   1. Olfactory cells. (D)
   2. Mucous gland. (C)
   3. Olfactory epithelium. (F)
   4. Olfactory tract. (B)
   5. Olfactory bulb. (A)
   6. Turbinate bones. (E)

2.
   (a) Taste bud.
   (b) Sweet.
   (c) Lingual, chorda tympani, facial or glossopharyngeal nerve.

3.
   1. Vitreous humour. (H)
   2. Optic nerve. (I)
   3. Aqueous humour. (B)
   4. Cornea. (A)
   5. Iris. (C)
   6. Sclera. (F)
   7. Ciliary body. (E)
   8. Suspensory ligament. (D)
   9. Choroid. (G)
   10. Lens. (J)

ANSWERS TO POST TEST

4.   A. Inferior oblique.

    B. Medial rectus.

    C. Superior oblique.

    D. Inferior rectus.

5.   (a) In speaking of the eyes' 'humour' we retain the historical meaning of this word — fluid.

    (b) Aqueous humour fills the anterior chamber of the eye and jelloid vitreous humour fills the posterior chamber.

6.   All three processes occur in accommodation: the axes converge in order to focus an object or objects that are near; the lens bulges in order to maintain focus; and the pupil constricts in order to cope with bright light.

7.   (a) The membraneous labyrinth.

    (b) 1. Saccule.   (C)

           2. Utricle.   (E)

           3. Cochlear duct.   (D)

           4. Ampulla.   (B)

           5. Semicircular duct.   (A)

ANSWERS TO POST TEST

8. (d) The auricle.

(e) The external auditory meatus.

(b) The ear drum.

(a) The oval window.

(c) The cochlea.

9.

	Rotation	Position	Sound
(a) Tectorial membrane.	( )	( )	(√)
(b) Macula of utricle.	( )	(√)	( )
(c) Semicircular ducts.	(√)	( )	( )
(d) Otoliths.	( )	(√)	( )
(e) Basilar membrane.	( )	( )	(√)

10. 1. Pacinian corpuscle. (C)

2. Golgi organ. (B)

3. Muscle spindle. (A)

4. In skeletal muscle. (A)

5. In tendons. (B)

6. Around joints. (C)